WILLIAM LEISS, author of *The Domination of Nature* and *The Limits to Satisfaction*, is with the Faculty of Environmental Studies at York University.

Politics is the forum where contending social forces are expressed, suppressed, or distorted; ecology, as the scientific understanding of environmental problems, now imposes its own vague imperative on the political contest. The issues are critical, but the answers are still far from clear. This book takes up the kinds of questions, the answers to which will shape the future. How do the pressures of social and political controversy affect our willingness to take action on environmental problems? How will our increasing understanding of problems such as pollution and occupational hazards affect the political trade-offs to which we are accustomed?

Twelve original essays focus on how ecology fits in with politics, current government activity, and received economic theory; on governmental inaction on questions of industrial and occupational diseases; on the social and environmental issues raised by our search for energy, particularly nuclear electricity; and on two procedural or administrative aspects of environmental policy-planning and environmental assessment. The editor concludes with an essay on 'Political Aspects of Environmental Issues.'

The book was organized under sponsorship of the University League for Social Reform.

D0887330

Ecology versus Politics in Canada

edited by William Leiss
for the University League for Social Reform

University of Toronto Press
TORONTO   BUFFALO   LONDON

© University of Toronto Press 1979
Toronto   Buffalo   London
Printed in Canada

Canadian Cataloguing in Publication Data

Main entry under title:
Ecology versus politics in Canada

ISBN 0-8020-2298-7 bd. ISBN 0-8020-6332-2 pa.

1. Environmental policy – Canada – Addresses,
essays, lectures. I. Leiss, William, 1939–
II. University League for Social Reform.

HC120.E5E26   301.31'0971   C78-001566-5

Foreword

The powerful social and political implications of many issues that pre-occupy scholars often lie unexplored both because the obscurity of academic publishing prevents communication with the public and because specialists typically have no opportunity within their profession to explore the general significance of their work by creatively confronting the views of experts in other disciplines. By commissioning books on problems that have been exposed to inadequate radical analysis, the University League for Social Reform aims to promote that intellectual fusion among scholars from different disciplines and also to translate their thinking into texts that are relevant and accessible to the interested public. Nationalism, foreign policy, urban reform, native land claims are among the topics on which we have already harnessed together intellectual talents from many diverse fields of competence.

This, our tenth volume, represents an important addition to our list, for William Leiss has managed to bridge two worlds – that of the ecologist concerned with environmental processes and that of the social scientist interested in government policy. This work in 'political ecology' has not been easy. It has taken four years from the conception to the birth of the book – a gestation that included a conference at which potential contributors could present their work for the scrutiny of their colleagues and the long editorial process of working with the authors on the second, third, and fourth drafts of the manuscripts that make up the chapters of this book.

As sponsor of this volume, the ULSR congratulates Bill Leiss for the skill, tenacity, and imagination with which he has brought this effort to a successful conclusion.

Stephen Clarkson
President
University League for Social Reform

Contents

Preface

Politics and ecology? *Political ecology?* The association of the two terms may seem strange at first glance, as strange as the phrase *political economy* must have seemed to our forbears two centuries ago. This collection of essays seeks to show how associating politics and ecology can help us understand some of the difficult issues our society faces today.

'Politics' has long had a double meaning, embracing both practical activities such as electing or overthrowing governments and the theories that suggest how we ought to organize our social bonds. The term 'ecology' often refers only to a special branch of the natural sciences, but during the last decade it has also been used more generally to identify the interests of groups who wish to change the ways in which we relate to our physical environment and the other living beings that inhabit it.

The nobler aspects of political thought and activity reflect attempts by men and women to regulate their public affairs according to principles which have, or at least are capable of having, a reasoned justification. Politics is the arena in which all the customary and institutional forms of interpersonal association are knit together, culminating in a general conception of 'legitimate' authority. From this conception is derived the sense of justice, freedom, and obligation that sets the bounds of individual action.

Ecology is also concerned with 'lawful' relationships. As a science it seeks to discern patterns of interactions that express relations between individual species and their environments and among species in specific parts of the environment (ecosystems). The actors in these settings may or may not include human agents, although by now it is difficult to find any part of the earth's biosphere that has not been affected, directly or indirectly, by human activity. And it is just this pervasiveness of human activity, this ineluctable pressure of human needs and numbers on every facet of the planet's biosphere, that has forged a bond between the concerns of politics and ecology.

Given the vast technical capacity of man to modify his habitat, and the power of nations and peoples to employ that capacity, there is today no major political decision that does not impinge on the functioning of complex ecological systems. And, conversely, there are no major ecological functions that do not have momentous significance for the fate of large human groups, if not for the whole species. This is the *practical* bond between politics and ecology: politics is part of ecology, ecology is part of politics.

True, man has modified his environment, and has been affected in turn by those modifications, even before the development of agriculture. But the cumulative impact eventually results in qualitative change: the modern industrial, chemical, medical, and nuclear-physics 'revolutions,' together with the increase of human numbers and the creation of a social 'world system,' which send perturbations originating in any sector rippling through the totality, make every new step more fateful and more uncertain. The stakes in this game have been bid up almost beyond calculation; if we think of something like the radioactive wastes from a large-scale nuclear generation program, it is increasingly uncertain whether the winners will have anything worthwhile to take home, or any home worth having.

Problems so vast call into question at once the capacity – and the credibility – of anyone so bold as to address them. The credibility of the authors represented in this collection will be enhanced, I hope, if the limitations they have imposed on themselves are stated at the outset. The over-all intention of this collection is to illustrate the link between politics and ecology through a series of separate but interrelated studies. Thus no claim is made to have presented a complete treatment of this theme. Second, the contributors have taken Canadian society as their focus; although some comparative judgments are unavoidable, the problems discussed are for the most part a sample of how the larger setting mentioned above has affected Canada, and how Canadian society has responded so far.

Thus these essays are, in the main, case studies. Broad conceptual themes are tackled in some of them; others deal with everyday situations where citizens now confront pollution or occupational health hazards. Some of the difficulties identified are so deeply entrenched in our established modes of behaviour and thought that no solution suggests itself at all. Others, for example some environmentally caused health problems, are amenable to great improvement in the short run, if sufficient importance were to be attached to them.

The two basic questions addressed in this collection are: How does social and political controversy affect our perception of, and our willingness to take effective action on, environmental problems? And how will our increasing understanding of environmental problems such as pollution, occupational hazards, and resource demands affect the political trade-offs to which we are accustomed?

The essays take up aspects of these questions in the following order. In the first three contributions a broad range of issues in two areas – ecology and politics, environment and economics – is identified. The next three move to a concrete level of controversy within those broader issues: environmental health in residential and occupational settings. At this level we can see how Canadian institutions have responded (or failed to respond) to what are only the first of a series of challenges to established policies in the zone where politics and ecology meet. Then this institutional response is discussed in depth, with public health and energy questions used as case studies; and this sets the stage for a more detailed look, in the two succeeding papers, at what is becoming one of the most contentious issues in Canada today: energy policy. The next two essays present complementary perspectives on other Canadian government initiatives, one in regional planning and the other in environmental impact assessment, and the latter includes a comparison with similar developments in the United States. The concluding paper attempts to present a unified statement of the different linkages between politics and ecology that have become known to us during the last few years.

W.L.
Toronto
August 1978

Acknowledgments

This book is the result of three different collaborative efforts, involving editor and contributors, editor and sponsoring group, and editor and publisher. As the go-between for all of them I am now well aware of both the pleasures and the hazards such an enterprise presents. Fortunately, the former outweighed the latter by an appreciable margin.

The contributors endured for more than three years a series of editorial memoranda and suggestions. Undoubtedly only a small portion of this interference was helpful; the complaints from those whose patience was tried beyond endurance, I am happy to report, respected the conventions of civilized decorum. The essays were all written specifically for this collection; I have learned that commissioning essays for publication can put a considerable strain on actual and potential friendships, and I can make only a most inadequate apology here for my deficiencies in academic diplomacy. I would like to express my gratitude to all those who responded to my invitation to participate in this project.

The executive of the University League for Social Reform agreed to sponsor this venture on the basis of little else save a good-natured faith in the editor's judgment. For this, and for the generous personal support that nursed the idea through difficult times, I offer my warmest thanks to Stephen Clarkson, Abe Rotstein, David Shugarman, Jack McLeod, Alkis Kontos, and their associates. The University League provided financial support for all stages of the project; beyond this, and of far greater importance, it has provided an unusually pleasant setting for serious discussions of new ideas.

Among the most welcome aspects of the project was the opportunity to continue my association with the University of Toronto Press. Rik Davidson and Hilary Marshall have a professional competence which is made all the more effective by their unobtrusive exercise of it. But it is in being welcomed to the lunchtime rendezvous with them and their colleagues on the Press staff where one realizes, with great pleasure and indeed with a sense of relief, that genuine fellowship and good conversation in the old style can still survive against the bureaucratic division of labour. The reader for the Press offered good advice, and Gerry Hallowell's copyediting turned a collection of essays into a readable book.

The Faculty of Environmental Studies at York University hosted a conference in May 1976 at which first drafts of papers were presented and discussed. I am sure that the contributors benefited as much as I did in being able to find the common ground that we sought in order to integrate our several concerns. Gerry Carrothers, then dean of the faculty, together with students and support staff, gave the generous assistance that allowed the conference proceedings to run so smoothly. Special thanks are due to Paula Engels for assistance with the manuscript preparation. The Canada Council, the University League for Social Reform, and York University provided funding for the conference.

Grahame Beakhust and I have taught a 'Politics of Environment' course together for three years, and I have learned a great deal from his lucid and provocative lectures. I hope that neither he nor Bob Sass, Peter Victor, Rob Macdonald, John Robinson, and other colleagues who have tried to instruct me will be too discouraged by the meagre results to date.

For the third time Marilyn Lawrence suffered with good grace the domestic consequences of book production, for which, so far as I am aware, no adequate recompense has yet been devised.

W.L.

Ecology versus Politics in Canada

1
Ecology and governments in Canada

Mark E. Taylor

People appear to be more aware of environmental problems these days and are constantly exposed by the media to pollution, food shortages, energy crises, and natural disasters. We are taking a greater interest in our environment and hardly a week goes by without some conflict between environmental groups and some organ of government; both the conflict and the organ would have been unheard of twenty years ago. Canada is a huge country composed of many different ethnic groups living in many different environments. It is consequently difficult to provide examples of all the various ecological/political interactions that currently occur. Most of my examples here will be from Ontario and British Columbia, since they are the provinces in which I have lived. The examples are only meant to illustrate my argument, so the list is not exhaustive; I hope therefore that readers in other provinces will be able to think of local situations similar to those mentioned.

Ontario is the major industrial province in the country, in that manufacturing and processing tend to dominate the economy, but it is also very important agriculturally and as a source of natural products such as lumber, oil, fish, and furs. Economists and politicians are continuously providing people with facts about productivity, the GNP, the level of unemployment, etc. These are tangible figures (even if seasonally adjusted) that we can put a figure or value on. The natural environment, however, is not so easy to measure and quantify. What is the value of an 'unused' clean stream as opposed to a 'used' polluted stream? How can we put a value on seeing cardinals or hummingbirds at the bird feeder or hearing geese as they honk their way north on migration? What is the value of catching your first brook trout of the season, or sitting round a camp-fire by a lake listening to the loons crying?

Unfortunately, things are changing; loons keep away from lakes frequented by power boats; song-birds fly into radio towers or die from applications of pesticides and herbicides. Since these changes and deaths do not have a dollar value, we are unable to measure the dollar depreciation of our environment as we might measure the depreciation of a building.

The provincial and federal governments are willing to extract various royalties and stumpage charges for the 'use' of resources. But are they really trying to protect the natural environment against the abrasive exploitation of man? A good example is the Ontario Water Resources Act of 1956, which was introduced to protect Ontario's water resources. Property-owners in two of the province's towns, Richmond Hill and Woodstock, suffering from untreated sewage flowing in streams through their land, sued the municipalities that operated the sewage treatment plants and won their cases. Under the terms of the act, the plants should have been closed down to protect the health of the towns' residents; instead, the Ontario legislature responded by passing legislation curtailing the rights of downstream land-owners! More recently, the federal government passed the Fisheries Act, another important piece of legislation for combating water pollution and protecting the aquatic environment. But the Canadian government delegated the responsiblity for enforcing the act to the provinces and recently it has been Ontario's policy usually not to enforce federal legislation.

In British Columbia there are also difficulties with the overlapping jurisdictions of various levels of government. Take for instance the case of the Afton Copper Mine and Smelter constructed between 1976 and 1978 near Kamloops. The provincial government decided to support the mine by providing a subsidy for the copper produced. The municipal and regional governments became involved since they were interested in the effects of the mine on the region and the city and because they have taxation control over it. The regional government was responsible for rezoning the land from agricultural to industrial use, but the city wanted to expand its boundaries to include the mine site and therefore have more control over the operation which will affect the city more than the region. In the spring of 1976 the director of the Pollution Control Branch issued permits to Afton without holding any public hearings, and construction of the site proceeded quickly. It was then up to private groups in the city to organize and appeal the pollution permits. As a result a hearing was held. By this time, the federal Department of the Environment had become involved since it was possible that the mine would contaminate streams leading into the Thompson River system. The result of the hearing was to uphold the permits, but to introduce certain new conditions concerning environmental studies and the control of mercury emissions. Also, the BC Pollution Control Board (composed of eight professionals from outside the public service and the assistant minister) instructed the director of the Pollution Control Branch that any proposed changes in the mine and smelter be referred to the board, instead of handling them himself. This represented an advance, since more public input will be possible in environmental decisions.

That concerned citizens had to raise money to appeal a decision made by a public employee and his publicly funded lawyer is a ridiculous perversion of democracy, but such conduct and events are but one of the absurdities that occur when governmental organizations are involved in environmental decisions.

In the organization of government lie some of the problems of relating ecological concerns to political decisions. Can it be determined easily whether it is the federal, provincial, regional, or municipal government which is responsible for preserving the environment for future generations? The fact that many environmental issues are taken to court and require that process for their resolution indicates that environmental responsibility is not clearly defined or enforced within the current political structure.

THE NATURE OF ECOLOGY

Ecology is a relatively recent branch of biology, though early Greeks, such as Hippocrates and Aristotle, wrote about matters that are now considered ecological. Much early ecological information was classed as natural history and the transition from natural history to ecology is still in progress. The main distinction between the two is that natural historians for the most part were content to observe and record natural phenomena, while the ecologist goes on to analyse and quantify the findings in order to determine the principles governing the natural environment. The science of ecology dates back to the beginning of this century, when it was considered to be the study of the structure and function of nature. Ecology is defined (in *Webster's Collegiate Dictionary*) as 'a branch of science concerned with interrelationships of organisms and their environments' or 'the totality or pattern of relationships between organisms and their environment.' There is no special preoccupation with humans here, for ecology is concerned with the totality of species, not just with one. So ecological considerations involve not only the particularly useful plants and animals and how they can best be managed for man's use or for aesthetic pleasures; they are about the relationships between *all* organisms, including man, and their physical environment.[1] Nor is ecology particularly related to problems of pollution, though pollution biology has unfortunately become an important part of ecology. Being a relatively young science, its laws and hypotheses are few and for the most part still being formulated and tested; and similarly, for most ecosystems, there is little knowledge about the many biological and physical components and how they interact.

This suggests one of the problems of our time. We know so little about the environment that, when we start modifying it, time and time again we blunder and are quite unaware of the problems we are creating. It is only in the present century that mankind has developed the technology to alter the biosphere significantly. Previously, our impact tended to be small and local; but now our pollutants – DDT, PCB s, CO_2, SO_2, oil, and many others – are distributed throughout the biosphere, uncontained by national boundaries.[2] In Canada, our ignorance about such basics as the distribution of our flora and fauna is more pronounced than in most other industrialized Western countries. There has been no systematic accumulation of knowledge of our flora and fauna comparable, for example, to the efforts by the Geological Survey of Canada to map the geology of the country. Admittedly, the country is 'new' (to the European settlers) and the population small per unit area, even in the southern strip; but this does not necessarily mean that we can afford to let it expand without due caution. We have no idea how large the human population could or should be on a long-term self-sustaining basis.

One of the principles of ecology is that a more diverse ecosystem is more stable, or better able to regulate itself within certain limits, than a simpler one, though this principle is not a law and has been subject to criticism.[3] Most of our agricultural processes involve a simplification of natural ecosystems, so that monocultures of wheat, corn, or vegetables are grown. This makes them far more vulnerable to pests.

Our approach to forestry is similar. It is more economical to harvest 'nice' uniform trees, preferably all the same species, than to have to log and process many species of trees of many different sizes. The spruce budworm, which affects the fir and spruce trees of eastern Canada, has a profound effect on the species composition of natural forests. In New Brunswick attempts to control the budworm by spraying have been undertaken for many years. Initially DDT was used, but in more recent years it has been replaced by Fenetrothion and other short-lived pesticides. The results in terms of control of the budworm have not been encouraging and barely justify the expense of aerial spraying though the forests remain green. But the way in which timber licences are granted by the government and the 'need' for logging companies to fell timber systematically is not reconcilable with the natural outbreaks of the spruce budworm that have occurred for centuries. The trouble with these outbreaks is that after the firs and spruces have been eaten and die, they are replaced by early successional trees such as aspen and birch which are not used commercially in New Brunswick. The spruce budworm thus maintains a natural diversity of trees in the forest, but this is not what logging companies like for profitable 'harvesting.' The spraying may also have side-effects on people living in the vicinity of the forests, such as Reye's syndrome which has been observed in several children in New Brunswick.[4] The question of whether or not to spray spruce budworm in the Fraser Canyon in British Columbia has certainly been controversial during the last year with the BC forest minister being forced to do a volte-face and cancel spray programs for 1978.

Invariably, when discussions of diversity occur, the question arises about the local or total extinction of species. Arguments continue about whether it matters if a few bald eagles or grizzlies or whooping cranes survive. The ecologist will usually answer that there are a number of reasons why species should not be forced prematurely into extinction other than for certain ethical reasons which I do not propose to consider at any length here.[5] There are many species which, although we do not know much about them and we do not 'use' them at the present, may be useful in the future. Once a species is pushed into extinction its gene pool is lost and there is no way it can ever be recovered. Much concern is being expressed these days about maintaining gene pools for plant and animal species, and the Man and the Biosphere program has resulted in the establishment of many reserves. Plant breeding is of critical importance to mankind, particularly since we appear to be reaching the limits of food production by conventional means. Many strains of plants and animals become weakened by continuous inbreeding and a great deal of hybrid vigour can be introduced by crossing the domesticated species back to the original variety. However, we do not appear, in Canada as elsewhere, to be maintaining very large reserves of potential breeding material; it is hard to find natural grasslands that have not been altered in some way by overgrazing or range management, for many of the grass species found in Canada have been introduced from Europe.

The productivity of marginal lands subjected to grazing and browsing by deer, moose, and caribou is probably higher than with cattle. Unfortunately, it is very difficult to evaluate all the parameters. If, for instance, cattle on marginal land eat grasses rather than shrubs and trees and the ground cover is consequently reduced, the rate of erosion will increase. How does one measure the consequent decrease in the value of the marginal land for grazing? Much of western Canada is heavily grazed at present and, although forest rangeland may appear to the tourist as natural and untouched, it is nevertheless a modified habitat and we do not know how long it will last before the soil becomes eroded. In the east, cattle manure produced on farmland rapidly breaks down and the land is fertilized; however, west of the Rockies the organisms necessary for the breakdown of cow pats are not present, so they dry out and do not yield their nutrients to the soil. This illustrates another facet of ecology. One cannot necessarily tell what will happen when new factors are introduced into an ecosystem on the basis of data acquired elsewhere. Studies usually have to be done locally to determine the important variables.

This then is one of the major points I wish to bring out. In order to have any hope of accurately predicting what will happen when an ecosystem is perturbed one requires a detailed knowledge of the various limiting factors and the components of the ecosystem in question. The acquisition of this knowledge takes a considerable amount of time, at least several years. Companies and governments wish to see action within their own short time horizons, and so want projects to be pushed through quickly. Thus proper environmental studies are infrequently performed. The requirements of knowledge and action do not coincide; the ecologist can on principle rarely agree with the time constraints imposed by a company which must produce an environmental impact statement of some sort as soon as possible.

It is very difficult to study natural ecosystems in southern Canada as there are only a few areas that are relatively 'virgin.' There are many reasons for maintaining natural ecosystems: as models for studying how natural systems work; as benchmarks against which we can measure the changes of our agricultural ecosystems; for their aesthetic value; and because interference in an ecosystem that is not understood may well be detrimental to it. It seems proper that some of our natural heritage should be preserved, but at the moment there are few exclusive areas where animals and plants can interact with their environment, fully protected from logging, mining, agricultural, educational, or recreational interests.

A good example illustrating the necessity for long-term studies and thoughtful planning occurred in Europe in 1976, when an unusually long hot drought affected many countries. Reservoirs, lakes, and rivers began to dry up and water restrictions were imposed. It was a 'freak' year, it was atypical, but what happened? The vast majority of domestic and industrial wastes in Europe are disposed of by water systems; pollutants are treated and diluted to concentrations that are not harmful to the majority of aquatic organisms. But with the long drought, the dilution of wastes decreased, the water became hotter thus carrying less dissolved oxygen, and hundreds of thousands of tons of fish died and had to be scraped up with bulldozers. Presumably it will be a while before the affected bodies of water recover.

What does this tell us? Basically, that many of our activities are rather too close to the limiting factors that control life, and that we only have to have some slight shifts in weather patterns to cause catastrophes.[6]

If there were more people and corporations living more harmoniously with their environment, that is, not throwing away so many noxious waste products, disasters would be less likely to occur. Are governments responsible, or is each and every individual? Is the government going to impose regulations as to what is a healthy, safe population living at a specified standard of living, or are populations and people going to be regulated by natural disasters?

THE NATURE OF ENVIRONMENTAL POLITICS

Politics is a very complex art or science, or mixture of the two, whatever one's perspective. It is symbolized in most people's minds by the huge maze of government offices in Ottawa and the provincial capitals, and all the municipal offices across the country.

There are three or four levels of government in any area depending upon whether a regional level of government has been inserted between the municipal and provincial. For many people, these levels of organization provide a bureaucratic barrier to any type of local involvement in planning, understanding local problems, or getting satisfaction in environmental issues.[7] Governments have also managed to confound people by changing, reorganizing, and abolishing departments or ministries. Thus the old Ontario Department of Lands and Forests has joined with Mines and Parks to form the large Ministry of Natural Resources. The Ontario Ministry of the Environment looks after some aspects of airports, water and air pollution, and so on, while a federal department, now called Environment Canada, includes the Canadian Wildlife Service and Forestry Service. The Ontario Conservation Authority has changed direction from its original responsibilities for managing watersheds and preventing floods by the judicious use of dams, reservoirs, and culverts to an attempt to provide parks and recreational areas where local authorities have been negligent or without funds. Agencies such as the Ontario Parkway Planning or the Niagara Escarpment Commission, also concerned with environmental matters, come under the ubiquitous Ministry of Treasury, Economic, and Intergovernmental Affairs. This reveals an interesting aspect of the Ontario government: the many boards, agencies, and commissions that have varying degrees of autonomy and may be responsible to no one but themselves.

To many people the Ontario Hydro Commission is such a creature. It has tremendous power to affect the environment – to dam rivers, string high voltage lines across the province, heat up rivers and lakes with thermal discharges, and release radioactivity into the atmosphere. In spite of this, it is relatively free to pursue its own goals, like selling electricity to the United States, without much governmental control. One could argue that controls do exist. The Solandt Enquiry was set up to investigate alternative rights of way for a proposed Hydro line across the southern part of the province. This gave people the option to intervene and put forward suggestions. However, this enquiry was not given terms of reference by which it could question the need for such power lines. Thus the Hydro Commission was able to sidestep the really important issue and provide less important ones for the people to discuss and argue about.

Factors such as the ethnic background of the inhabitants, the amount of industrialization, and the degree of natural resource exploitation combine to influence the atmosphere of the provinces. However, there are weaknesses in the present municipal structure and one of them is the inability of most of them to finance even modest improvements in the quality of their surroundings. Their main income is drawn from property taxes which are usually insufficient to meet the needs of growing communities where, besides roads, sewers, schools, libraries, and hospital facilities, there are also needs for parks and recreational areas. It is also easy to spend a large proportion of the parks and recreation budget without much to show for it. For instance, Richmond Hill, a town of 35,000 inhabitants just north of Metro Toronto, decided to allocate a large part of its 1975 budget to renovate the Mill Pond which was silting up from additional run-off from streets in that part of town. Most of the money will go to an engineering company to drain the pond, and move the offending mud to an upstream swampy area. This is probably not the wisest use of the available funds, considering the general facilities of the region; the nearest conservation areas are ten miles away and require people to use cars to get there.

It is also questionable whether municipalities can afford to bother about protecting the environment and preserving interesting local ecological areas. A marshy area that is an important stopping place for migratory waterfowl may also have potential value for building, and it might well seem absurd for a town council to protect a mosquito-infested swampy area close to town. However, if all small towns and municipalities were to drain their ponds and marshes, the effect on the migrating waterfowl population would surely be disastrous.

Provincial governments recognize problems of this nature and some have already introduced regional governments to try and reorganize the local government structure. The goal behind regional development and planning is to provide each region with a fair share of the process of growth and change and systematically to guide the destiny of an area with common characteristics. These regional governments should be able to plan far more comprehensively than the separate towns, and by having a wider resource base be able to afford competent experts in the various fields necessary for co-ordinating such an operation. The proposed population for such regions is between 150,000 and 200,000 inhabitants. However, there are problems associated with setting up such regions. The Ontario Tory government tried to establish one in the Hamilton-Wentworth area to provide what it considered a balanced region. But the proposals met with stiff resistance from a number of citizen groups and the government eventually modified its plans considerably in order to make the new region acceptable to the voters. It is clear that the provincial government does not wish to introduce a system of regional government that would antagonize people and bring about its defeat at a general election. This introduces a difficult political problem. Should the government assume it knows best, behave autocratically, and impose decisions – or should people's immediate interests be satisfied and the future left to look after itself?

The Ontario government is certainly aware of environmental problems and concerns. D'Arcy McKeough said back in 1972: 'One of the major and insistent issues of our time is the quality of life and conservation and preservation of our environment. Taken as a whole our "system" of local government is unsuited to providing broad policies and priorities to tackle this issue.' Unfortunately, there does not appear to have been much progress since then.

One of the important questions seems to be whether or not the Ontario government or any other democratic government can successfully plan and guide a large area so that all human interests are entertained and the natural environment is not over-exploited and gradually destroyed.

ENVIRONMENTAL LEGISLATION

Since the International Biological Programme of the 1960s there has been a definite move towards introducing legislation to protect and conserve representative parts of ecosystems all over the world. In Canada two provinces, British Columbia and Quebec, have moved ahead in this area. The Ecological Reserves Act (1972) of Quebec is the more idealistic and preferable piece of legislation. In stating its objectives it says: 'The purposes of the establishment of Ecological Reserves are:
1 / to protect representative samples of most if not all natural ecosystems;
2 / to establish territories of interest for scientific research and education where it is possible for scientists to study natural sequences and other aspects of the natural environment;
3 / to establish gene pool reserves and protect threatened, endangered vanishing animal and plant species;
4 / to reserve suitable territories for the study of geological, geomorphological, pedological, and ecodynamic scientific phenomena;
5 / to reserve territories suitable for scientific study of different environmental parameters for monitoring purposes;
6 / to permit in the buffer zone as defined in section 1e, the establishment of Ecological Centres with educational purposes, in accordance with the management plan of a reserve.' (Buffer zone means a territory included within the limits of an ecological reserve with the purpose of reinforcing the protection of the reserved zone and where activities dedicated to education and scientific research can take place according to the management plan.)

The British Columbia act does not delineate a central core and buffer zone, as does the Quebec act, nor does it prohibit entry of people into the reserve as the Quebec act also does (unless permission from the minister of lands and forests has been obtained). However, since the British Columbia Ecological Reserves Act was the first such act introduced in Canada, it had to be, to a certain extent, experimental. Since it was given approval by the legislature in 1971, eighty-eight reserves have been gazetted with a total of 88,904 hectares. They range in size from 0.61

hectares to 33,200 hectares (average 1010 hectares) and range from subtidal marine parks at the coast to high alpine habitats.[8] Each reserve has been set aside to preserve one or more natural areas from interference by man, and the number of reserves is steadily increasing. These ecological reserves in Quebec and British Columbia are specifically for the protection of natural areas and are not to be confused in any way with provincial and national parks which are for the most part recreationally oriented.

Ontario does not have any ecological reserves, its closest approximation being the large provincial parks. There are three of these: Polar Bear, with 5,952,000 acres; Quetico, 1,150,404 acres; and Algonquin, 1,862,400 acres. Neither Polar Bear nor Quetico is subjected to use by recreationists in the way that the more southern parks are, and so they might be considered as a type of ecological reserve. Algonquin is being used more and more for recreation and with the public pressure for more campsites and canoe routes very little of the park is now inaccessible to urban canoeing man. It is also still being logged extensively – a contentious issue with many environmental groups.

There are no large protected areas in the southern part of the province – one region that is particularly distinctive to Ontario. Many relatively exotic species are found here that are not found to any extent in the other provinces. Natural areas such as Point Pelee and Inverhuron should well be expanded on ecological grounds; in fact, however, Inverhuron was closed down to allow the expansion of Ontario Hydro's new atomic reactor and heavy water plant. There are also a large number of conservation areas in the province, but for the most part these are primarily organized for recreational activities. The Conservation Council of Ontario may acquire land of interest for the public and has a small budget with which it is able to make immediate offers on a property when it comes up for sale. The properties can then be transferred to the government and the council reimbursed to make further acquisitions. In this way a number of valuable areas of ecologic interest have been saved from the developer's bulldozer.

When one considers that many Ontario parks, such as Algonquin, are multiple-use parks where activities such as logging, trapping, and fishing are permitted, they bear even less resemblance to ecological reserves and they cannot be considered as reasonable alternatives. The closest approach to an ecological reserve act created by the Ontario government is the Wilderness Areas Act. This, however, is a totally misleading title for it only protects areas up to one square mile (640 acres) which, excluding a few exceptions, most ecologists would consider to be too small for a self-supporting ecosystem – let alone be a 'wilderness'! The act also comes under the management of the Parks Division of the Ministry of Natural Resources whose preoccupation is with recreation and tourism and is thus likely to produce internal conflicts of interest. From the way in which the Wildlife Research Area, used for over twenty-five years in Algonquin Park, was abandoned in the recent master plan, it is plain that political decisions could easily remove any wilderness areas or ecological areas unless sufficient precautions are taken. In the Quebec Ecological Reserves Act the supervisory committee is composed of a majority of people outside the public service who are given the power to hold public hearings about proposed changes in gazetted reserves, thus protecting the reserves from changing government pressures. The general concept of ecological areas is that they must remain so more or less in perpetuity and not be subject to political whims.

This again illustrates the problems inherent in protecting and maintaining viable ecosystems over long periods of time and of utilizing 'all available resources', as governments are pressured to do to satisfy 'immediate needs.' One of the (potentially) more important government acts protecting species is the federal Migratory Birds Convention Act and Regulations, now controlled by Environment Canada. Protected species are listed and game species (such as ducks and geese) are specified along with their open seasons. The provincial governments are responsible for designating open seasons for the different species in the various regions of the provinces. However, the majority of migrating birds such as ducks and warblers are dependent upon groups of ponds or woodlands occurring along their flyways and, if these areas are 'developed' or destroyed, then the chances of migrating birds surviving are severely curtailed. Likewise large buildings and radio masts are responsible for the deaths of large numbers of 'protected' migratory birds, but the owners of such structures are not prosecuted under the terms of the act. Clearly the act is based upon good intentions, but the government does not have the means to enforce it. Since little data is available on the numbers of migrating birds in the country or on their mortality rates due to man-made obstacles, direct or indirect, it is impossible to decide how strongly the act should be enforced. Again the problem lies in our over-all lack of interest in and information about our ecosystems – and it is difficult to protect species when we know so little about them.

It would appear from reading the daily papers that the Ontario government is very actively involved in protecting the environment. Announcements about the Parkway Belt or the Niagara Escarpment make one believe that the government is indeed planning well for the future. An examination of the master plan for the Parkway Belt gives a very different impression. The Parkway Belt is to extend from Dundas in the west to Oshawa in the east and is composed of two parts: Parkway Belt West from Dundas to Hwy 48 at Markham and Parkway Belt East from Hwy 48 to Oshawa.[9] The objectives of the Parkway Belt system are:

1 / community identification,
2 / integration of two-tier system of urban areas,
3 / land reserve for future flexibility,
4 / linked open space framework.

None of the goals of the plan are related to preserving ecosystems, although it does suggest maintaining existing vegetation including hedgerows, woodlots, orchards, streams, ponds, and marshes *where possible*. But what happens if any of these features lie in the path of the proposed highways or power corridors? The plan is essentially proposing a belt of roadways, transmission lines, and other utilities to run in an arc from Dundas to Oshawa. Recreational green areas and parks are included in the plan, but their close proximity to a major freeway is rather odd, and the fact that most of them are already in existence does not suggest that the Ontario government is planning any new extensive recreational areas in the Toronto region.

The Niagara Escarpment Planning and Development Act passed in June 1973 also appears on the surface to be a well-meaning and effective act to protect the unique geological, historical, and ecological character of the escarpment from the gravel and other industries. It sets out to provide a north-south corridor across peninsular Ontario which would presumably provide a route for the movement of many animals. The act states that its aims are to protect unique ecologic and historic areas and to maintain and enhance the quality and character of natural streams and water supplies. However, the government is not planning to acquire any large parcels of land but intends to control 'development' within the boundaries of the escarpment region by placing it under the jurisdiction of the treasurer of Ontario and the minister of economics and intergovernmental affairs. No criteria for issuing permits are provided in the act; permission depends upon a set of regulations specified by the minister in charge. Since the minister is also allowed to delegate authority to the Niagara Escarpment Commission and to counties and regional municipalities, one wonders how well this act will be enforced, given the political considerations permitted to influence his discretion.

Steps are certainly being taken to protect some important ecological areas in southern Ontario. For instance, the plan for the Minesing Swamp prepared by Raymond Moriyama and various committees is a step towards analysing a reasonably large acreage of great ecological interest. The swamp covers some 15,000 acres and contains a number of fascinating habitats which during the year attract numerous rare migrants and provide a permanent home for many residents. The plan is for a multiple recreational use of the swamp with some areas closed off because of fragile habitats or breeding birds. It is to be hoped that the government will consider this as a 'reserve.' However, at the moment it will have to be legislated as a special ecological section of the Conservation Authorities Act under one of the other environmental acts. Again, without an ecological reserves act in the province, it is not easily decided under which jurisdiction a particular reserve might be placed. The Minesing Swamp could as easily be controlled under the Ontario Water Resources Act as under the Conservation Authorities Act. Likewise, hypothetically an aquatic ecological reserve in one of the Great Lakes could be controlled by the Ontario Water Resources Act or Beach Protection Act. Other than the various plans under consideration by the various conservation authorities and the Ministry of Natural Resources, there is no government activity to preserve the natural ecological heritage for future generations.

ECOLOGICAL INFORMATION

One of the basic problems that governments and other organizations face in making decisions related to the environment is lack of data. Except for some economically important species of animals and plants there is, as we have noted, very little data available on species distribution and abundance. How then can well-founded decisions be made? The answer to date must surely be that most decisions are rather arbitrary and based on small amounts of local information. There is no organization within the country to collate the existing data on the different species in the wide variety of ecosystems in the provinces and in the nation as a whole.

In Britain there has been a Biological Records Office since 1964 to map the distribution of the flora and fauna. This mapping endeavour started in 1954 for the preparation of distribution maps of the British flora. It was published as an *Atlas of the British Flora* in 1962. Subsequently, a number of organizations have set about preparing distribution maps for many different groups of organisms, from otters to insects, and some of these are already available. One result has been the implementation of legislation to protect otters and badgers as these two species were found to be endangered. Such regular mapping is also useful in determining how control measures on species such as the introduced nutria and mink are progressing. National biological record centres now exist in Belgium, Czechoslovakia, Finland, France, Germany, Italy, Luxembourg, Norway, Poland, Ireland, Sweden, and Yugoslavia. And the endeavours of each nation are co-ordinated so that the method of collecting and recording information is standardized; some day there will be an International European Biological Record Centre with access to reliable data on all the major animal and plant groups.

There are, however, no plans for organizing provincial or national inventories in Canada. It is only when important areas are documented that it will be possible to protect them, and that development could be guided into those areas which are ecologically less interesting or have already been seriously disrupted. But the question still remains as to how much of the land should be exploited or managed and how much left under relatively natural conditions? Since we do not know how many ecosystems work, and have a very poor record in manipulating them, it would seem foolish to try and exploit them all to the last dollar. These problems are of great importance to the well-being of man, as well as to all other organisms, and the provincial and national governments are probably the only organizations capable of organizing and financing research and planning at this level. It would seem appropriate that they should undertake to determine far more exactly the nature of the ecology of the country.

There is already a considerable amount of ecological research going on in the provinces by different groups and agencies. First of all, there is research in the universities by professors, graduate students, and technicians – a large-scale concern, though, for the most part, the groups tend to work separately. Much of the information obtained is more theoretical than practical but this is not to suggest that the work has no practical application – often it may be linked to government programs in the areas of wildlife, agriculture, and fisheries. As there is not yet a 'Canadian Journal of Ecology,' the information that is made public usually appears in such publications as the Canadian journals of zoology, botany, fisheries, and forestry. Papers are also published in American or European journals of ecology. The various government ministries that deal with ecological or environmental concerns also publish a considerable amount of information every year.

The Ontario Ministry of Natural Resources (and its equivalents in other provinces) publishes papers on specific topics in addition to maintaining a number of internal mimeographed reports dealing with the on-going work of various sections. The Parks Branch of this ministry prepares plans and inventories for different parks which may contain useful ecological information. Agencies or commissions of the government such as the Ontario Hydro Commission and more recently BC Hydro also employ a number of ecologists whose findings are not immediately available to the public but may be published sooner or later. Under the Niagara Escarpment Planning and Development Act provision is made for the protection of unique ecologic and historic areas together with the maintenance of the quality and character of natural streams and water supplies, and presumably some research will be done to substantiate the claim to these areas and to catalogue them.

Federal agencies are also involved in ecological concerns in the provinces through such departments as the Canadian Wildlife Service and the Forestry Service. More research has been done under international auspices, particularly under the International Biological Programme, the results of which are currently being published.[10] Environmental consultants also do ecological research for the government or private companies. Large international companies such as Dames and Moore or smaller national ones such as LGL carry out ecological surveys prior to some development, but generally the data acquired is private information and does not become public unless the client so desires.

The combined capabilities of the universities, provincial and federal governments, consulting firms, and naturalist clubs are enormous. Unfortunately, most of the groups tend to go their own way with the minimum of interaction, and for some there is even little interaction within the group. Provincial governments have regional biologists as well as research facilities distributed throughout the provinces. Again, it would seem a wise policy if these people were involved in the development of a national network of ecological data gathering and processing.

Such a policy would certainly help in the planning and development of the country. In the long run it would save time in planning developments because the information as to where new cities, industries, etc., could and could not go would already be available and there would not have to be environmental hearings for every new scheme that is proposed. It would involve large numbers of people with their local natural environment and would enable them to be more aware of their heritage. And, lastly, such a venture would encourage co-operation between the many diverse groups interested in the natural environment, and reduce the amount of time spent 'fighting' causes in environmental hearings.

NOTES

1 This was recently mentioned by V.B. Scheffer in the context of wildlife management. See his 'The Future of Wildlife Management,' *Wildlife Society Bulletin*, 4 (1976), 51–4

2 There have been a considerable number of scientific and general papers on the impact and potential impact of pollutants. For a recent discussion, see almost any issue of *Science* in 1977–78; also C.F. Baes *et al.*, 'Carbon Dioxide and Climate: The Uncontrolled Experiment,' in *American Scientist*, 65 (1977), 310–20; and W.R.P. Bourne and J.A. Bogan, 'Polychlorinated Biphenyls in North Atlantic Seabirds,' in *Marine Pollution Bulletin*, 3 (1972), 171–5.

3 D. Goodman, 'The Theory of Diversity-Stability Relationships in Ecology,' in *The Quarterly Review of Biology*, 50 (1975), 237–66

4 R. Hunt, 'The Human Cost of Spraying the Budworm,' *Weekend Magazine*, 7 Aug. 1976. Also there is another good account of the spruce budworm spray program in New Brunswick, showing some of the interactions between public concern, big business, ecological matters, and political decisions, in 'Spray Fever: The Fight Is On,' by R. Conlogue in the *Globe and Mail*, Toronto, 3 June 1978.

5 For a good review of this question, see R.L. Smith, 'Ecological Genesis of Endangered Species: The Philosophy of Preservation,' in *Annual Review of Ecology and Systematics*, 7 (1976), 33–55.

6 For a review of these limiting factors, see E.P. Odum, *Fundamentals of Ecology* (Philadelphia 1971).

7 This is clearly revealed by a study of some of the cases illustrated in D. Estrin and J. Swaigen, eds., *Environment on Trial*, Canadian Environmental Law Association and Canadian Environmental Law Research Foundation (Ottawa 1974).

8 'Ecological Reserves in British Columbia,' April 1978, British Columbia, Ministry of the Environment

9 'Interim Draft Parkway Belt West Plan.' May 1975, Ontario, Ministry of Treasury, Economics and Intergovernmental Affairs

10 For instance, a list of ecologically important sites for the Canadian north is listed in 'Ecological Sites in Northern Canada,' 1975, ed. D.N. Nettleship and P.A. Smith.

2
Political ecology

Grahame Beakhust

Political ecology is an idea and a movement. The movement, concerned with a multitude of issues from cultural destruction to nuclear energy,[1] contains all manner of dissenters from former cabinet ministers and middle-class neo-primitives to Marxian analysts and the occasional labour leader, with a heavy emphasis on intellectuals and academics. The idea, though relatively new in its present form, has lengthy intellectual antecedents, and it is from this source that a clear-headed analysis of the relationship between environmental degradation and social structure is most likely to emerge. That analysis is by definition political.

Aldous Huxley has argued that 'in politics, the central and fundamental problem is the problem of power,'[2] but this begs the question of the necessity of power and its exercise in human affairs. Politics is more fundamentally concerned with the question of how we might best live together, and as industrial society spreads the question increasingly becomes one of how we might best live together within the limits of our natural environment or, more simply, how we might best survive. Thus as biology comes to challenge the dominance of physics in the post-war world, political ecology emerges as a contemporary image through which the knowledge and understanding of art and science may be combined and reconciled.

The combination of politics and ecology is intended to highlight a duality of dependence rather than independence. The intimacy of this link is rarely lost on subsistence societies, whether they express it in formal social arrangements or highly developed cosmologies.[3] The purpose in using the idea here is to focus on the necessary connection between politics and ecology in a society whose industrial successes – or excesses – have led it to believe that nature as well as man can be subjugated by its political will. The connection is a dialectical one that transcends what Engels characterized as the 'hollow abstraction' of cause and effect and emphasizes rather that 'the whole vast process goes on in the form of interaction.' Political ecology is not, then, susceptible to the reductionist form of analysis common in much of the natural and far too much of the social sciences. Its thrust is legitimately general and integrative, as concerned with images in its grasp for intellectual understanding as it is in the call for political action. It draws upon the fact and analysis of our scholarship, but also upon the informal, craft skills of other cultures that 'have often displayed remarkable success in interpreting and coming to terms with their specific environments.'[4]

Politics is an ancient skill and even modern definitions talk about 'the science and art of government ... the science dealing with the form, organisation and administration of a state or part of one ... that branch of moral philosophy dealing with the state or social organism as a whole.'[5] It is, like philosophy, science, and art, among the earliest of all human activities.

According to Hans Enzensberger, the concept of ecology 'emerged for the first time in 1868 when the German biologist, Ernst Haeckel, in his *Natural History of Creation*, proposed giving this name to a sub-discipline of zoology – one which would investigate the totality of relationships between an animal species and its inorganic and organic environment.'[6] This was, as he suggests, a comparatively modest program 'when compared with the present state of ecology' and there is no doubt that the contemporary use of the term has expanded, perhaps to the chagrin of those among the natural scientists who would have it all to themselves.

Political ecology has a much better known alliterative relative in 'political economy.' The common 'eco-' derives from the Greek word *oikos*, meaning 'house' or 'family' or 'household'; *logos* refers to 'one who speaks in a certain way' or 'one who treats of a certain subject'; while *nomos* means 'law' or 'distribution.' Adam Smith defined political economy as 'the art of managing the resources of a people and of its government' and fifty years later John R. McCulloch described it as a theoretical 'science of the laws which regulate ... production, accumulation, distribution and consumption.' As both a theory and a practice, its crucial departure from the past and increasing importance through the nineteenth century lay in the commitment to capital accumulation or what is now conventionally called a 'growth economy.'

There is some danger in this rather simple characterization of political economy. *Nomos* carries the implication of just allocative regulation that is retained in Marx's humanitarian vision of political economy with its emphasis on distributive and class justice. This view may well be reconcilable with the idea of a necessary order of nature contained in the notion of *logos*. The vision of political economy to which I refer and which I am concerned to criticize is that of economic science, from which emerges the model of society as a mathematical jungle of competing, individualist 'interests.' The success of this now dominant model in delivering an ever-increasing and diversifying supply of commodities is unquestioned. Its weaknesses lie rather in the social and environmental spheres.

Within both the discipline and the reality, the natural environment is treated as a storehouse of raw materials and a convenient dump for the by-products of development. People are similarly treated, and valued, as human machines. In an early episode of *Roots*, the timeless TV torture test of 1977, a doctor responds to the deliberate maiming of the hero by noting the damnable waste of a man/machine that could otherwise have repaid handsomely the slave-owner's initial investment. In a more humane, but similar, vein, the Toronto *Globe and Mail* on 25 January 1977 quoted a missionary who said of the Masai that some of their herds 'are worth thousands of dollars, but the men just don't like to sell their cattle and the women need the cows for milk.' The point in both examples is that the economic science conception of our species is still dominant; it is a conception that treats people as 'labour' or as one among several 'factors of production.' Though causality is difficult to prove, there seems a close connection between this attitude towards people and the equally simple and savage perception of the planet as a conglomeration of 'resources' awaiting 'development.' The Toronto *Globe and Mail* of 9 August 1978 quotes Ontario's new royal commissioner on the northern environment, John Fahlgren, as stating: 'Believe me, God gave the natural resources to you for use. He didn't put them there for anything else. What we have to do is develop them and market them in the best way we can.'

Political ecology, by way of contrast, 'treats of a certain subject' – the relationships, in this case, between people and their environment – and to a degree 'speaks in a certain way' – that of an analytical concern about organic evolution, rather than a prescriptive model designed for the specific goal of economic growth. In *The Doomsday Syndrome*, John Maddox argues that 'ecology is no longer a scientific discipline ... it's an attitude of mind.'[7] Even if it were true, the observation is more worthy of celebration than lamentation. At the simplest level, political ecology redirects our attention from a politics of economic growth that treats the natural world as a passive non-entity towards one that takes seriously the scientific proposition that any species seeking survival must live in reasonable harmony with its environment. However one relates to this argument, its undeniable correlative is that we should no longer treat each other, or allow ourselves to be treated, as cantankerous fleshly machines whose principal purpose is to seek work and then do it.

There is a second meaning to economy that bears a closer relationship to ecology. While its roots are commonly translated as 'household management,' and extended as political economy to the expansion of the national household, the notion of economy as careful management and thrift was a common connotation as early as the seventeenth century and applied to minimizing outlays of labour, time, and money – 'resources' all.

By the turn of the last century resource shortages and price increases helped combine both concepts of economy in the early conservation movement in the United States. One of its 'great mainstays,' as Frank E. Smith noted, 'has been the deep-rooted philosophy that our natural resources must be developed in the interest of economic well-being.' The movement of this period was, as Carol Bailey has observed, 'primarily concerned with the wise and efficient use of resources to ensure industrial growth, not to inhibit it.'[8] Ironically, it represents a step away from the classical free-trade-for-capital arguments of early political economy, not in the direction of environmental protection but in order to protect the national resource base for domestic exploitation.

Sir Clifford Sifton's policies in Canada bore the same stamp, one that can still be seen today in such legislation as the Oil and Gas Production and Conservation Act, which also highlights the dual neglect of people and environment at work in the 'conservation' approach to economy. Speaking to the committee considering the bill in 1969, the deputy minister of Indian Affairs and Northern Development stated that the government's approach to questions of oil and gas was 'from the point of view of the system which will produce the lowest cost product.' John Rodman has characterized this early form of 'ecological consciousness' as 'the economic mode of prudence.'[9] A form of conservation designed to protect the environment not for intrinsic reasons but to promote future and efficient growth has significant effect on people as well. Mine shafts are driven and factories constructed with a view to maximizing profit, if not product. Scant attention is paid to geology, hydrology, or biology and none at all to subsequent impact and what happens when their useful lives are over. While nuclear plants will soon be the most glaring – and effectively everlasting – example, gold mines at Yellowknife are already leaking arsenic wastes into local drinking water, wastes that have been casually reinjected into worked-out shafts for the past three decades. Mine workers are likewise considered as part of the profit calculus, their conditions improved only when they result in poor productivity, or more recently when unions feel sufficiently strong, even given the threat of strikes by capital, to force change upon owners. The rights of this latter group are, however, much more scrupulously protected by the state. In the case of the act discussed above, the deputy minister stated that its philosophy was 'to secure the working co-operation of the industry under the law with latent compulsory provisions.' In referring to the power the bill granted 'to order the cessation of all operations giving rise to waste,' the assistant deputy minister was quick to add: 'Mind you, unless it is a case of threat of personal injury or to life, there has to be an investigation to protect the rights of the people who are alleged to be committing waste.'[10]

By way of contrast 'the thing to do' with regard to the environment 'is to keep the damage to a minimum.' Little attention was given to northern native people whose land and life were (and are) threatened by hydrocarbon development. They were typically viewed as labour for whom industry would provide 'openings' and opportunities for 'advancement.' The current political economy of conservation, then, requires the highest level of protection for the rights of property, a minimal protection of the environment to permit continued profitable production, and sufficient gestures towards the people affected to keep them quiet.

Political ecology is not about conservation in this limited sense. It does argue for a conserving approach to the natural world, but one based more on Rodman's later stages of ecological consciousness, namely 'the moral/legal mode of conscience, the religious/esthetic experience of transcendence through myth and ritual, and the immersed awareness of the metaphoric sensibility.' In both theory and practice political ecology transcends and incorporates much of political economy. It, too, is concerned with connections and equilibria, or, as Edgar Morin terms it, 'the great law of the ecosystemic relationship ... the dependence of independence.'[11] But it is in the factors it considers and the way it considers them that political ecology marks a new departure.

In an era when 'the economy has lurched from the promise of abundance to the old fact of scarcity ... the late shift of attention to cultural forms of liberation may have to readjust itself not only to an old focus on economics, but also to a new focus on nature.'[12] Nature must enter into any political calculus, but the exercise itself must focus on the central question of liberation since 'men pay for the increase of their power with alienation from that over which they exercise their power. Enlightenment behaves toward things as a dictator toward men.'[13] Alienation from one another is the price of social and economic domination; that much is clear from the Marxian critique of capitalist political economy. What must now be considered is the price paid for alienation from the natural environment, an alienation almost imperceptible in urbanized and industrialized society where concrete is the medium for growth and the natural is programmed in parks and playgrounds and television. Despite all that distances modern man from nature, they cannot fundamentally be separated. Mastery is the myth and alienation the reality, however carefully they may be concealed and contained. In Vincent Di Norcia's words: 'We must rewrite Lukacs' dictum that "nature is a social category ... whatever is held to be natural is socially conditioned." For human culture is itself a category of natural history; whatever a culture holds to be social for itself is powerfully shaped by its habitat.'[14] In turn, the method of political ecology, like that of animal ecology, must supersede the old categories of cause and effect and adopt, rather, one of co-evolution.

The earliest use that I can find of politics and ecology in combination is in a 1951 speech entitled 'The Political Ecology of Alaska,' by its then governor, Ernest Gruening. The principal subject of the speech was a demonstration of the need for statehood in Alaska drawn from its political history. His use of the term suggests it as a possible scientific approach to the understanding of the process of Alaska's economic development. The speech seems to have been relegated to the obscurity such offerings deserve, and the idea did not surface again for ten years, when it appeared in George W. Rogers' book, *The Future of Alaska*. Subtitled the 'Economic Consequences of Statehood,' it contains a section on 'The Political Ecology of Resource Development' in a chapter with the grand title of 'The Great Land.' The term appears three times. He notes that 'Any investigation of Alaska's natural resources and its prospects for future growth must become a case study in the political ecology of development' and again that 'the achievement of statehood represented a major change in the political environment and in the political ecology of Alaska's development.' The final reference is to 'the sharp change in Alaska's political ecology represented by statehood.'[15]

These statements are not on the whole very enlightening, and the first, like the general tenor of the book, is clearly set within the context of utilitarian resource conservation. It is unclear what distinction Rogers wants to draw between 'the political environment' and 'the political ecology of Alaska's development.' The term seems to have come to mind during a required re-reading of Gruening's speech, that was itself intended more as an inspirational headliner than the unveiling of a new conceptual analysis of Western civilization. The intent of highlighting the intimate relationship between ecology and development, or more specifically the northern environment and the development of Alaskan resources, is not surprising in a state where their extraction was running a close second to that of Pentagon payrolls in the local economy.

There are, however, important undertones in Rogers' analysis. At the most obvious level, he wants to focus on the convergence between the dominant environmental issue of resource development and the political processes involved in the statehood debates. At the same time he recognizes that the 'resource endowment' of a country or state, as a description of its development potential, cannot be arrived at simply by tallying up projected yields from each resource group. Timing and the sequence of developments also have a role to play in what Barry Weisberg calls 'planning for the efficient development and use of all natural resources as opposed to the robber baron spoil mentality.'[16] But beyond these essentially estate-management considerations, there is a whole range of what Rogers calls 'operative ideals' – the broad goals of the society itself. Is, for example, a resource endowment to be mined out in the interest of temporary advantage and short-run private gain, or managed so as to maximize total benefits, material or otherwise, for the entire society? Are local, or national, or international interests to make the key decisions about development? Though Rogers never states it explicitly, the natural extension of these ideas is to begin questioning from a non-utilitarian perspective, asking such questions as whether a society's goals are to be based on or realized through immediate material gain, or through a non-materialist, ethical concept of conservation in the sense of wise non-use.

Enzensberger's article, published in 1974, uses the term political ecology only once, in the title. A more recent reference, and apparently the first Canadian one, is Edgar Friedenberg's description of William Leiss' book *The Limits to Satisfaction* as 'a substantial contribution to the literature of political ecology.' The central proposition of Enzensberger's paper is that 'the social neutrality to which the ecological debate lays claim, having recourse as it does so to strategies derived from the evidence of the natural sciences, is a fiction.' Read in conjunction with David L. Sills' comprehensive, but less critical, overview entitled 'The Environmental Movement and Its Critics,' it offers an important analysis of the problem from the perspective of political ecology.[17] What both Enzensberger and Leiss are doing, one more politically, the other more philosophically, is highlighting and analysing the links between politics and ecology. Many authors make similar points[18] and the issue is summed up succinctly by Sills: 'Environmental policies both cope with and increase scarcities of various kinds; they encourage a certain type of lifestyle and discourage other types; they activate interest groups; they create conflicts over environmental decisions; and they both influence and are influenced by those who hold power in society. In short, they are political policies, and since the environmental movement seeks to influence the formulation and execution of these policies, it is in the broad sense of the term a political movement.'[19]

Political ecology offers an opportunity for bringing together the social and analytical aspects of this movement; it can help to analyse its social consequences while socializing and politicizing its analysis. A couple of examples might help. In 1977 the Public Broadcasting System produced a program on the coal-mining industry in the United States. One interview was with a mid-western rancher who argued forcefully, as I have heard many Inuit and Dene argue, that he was above all else a rancher, and wanted to stay a rancher despite the considerable inducements offered by the coal industry to get off his land and, quite literally, sell out. 'Being a rancher,' relating as a rancher to his natural environment and social milieu, rather than as a bovine tending primate, was the essence of his identity as a person. The same program also dealt with West Virginia. It showed the degradation of people going hand in hand with that of the land. It was a simple and graphic demonstration that even in economic terms the cost involved in developing cheap industrial resources are both environmental and human, in the direct sense as well as the derived sense in which environmental damage offends middle-class aesthetics. Black lung is the price people are forced to pay for a lifestyle substantially below the material average. It is one they do pay, as do the fluorspar miners of Newfoundland, not only because mining is all that they know, but because that is their land, their lives, and their history. I was struck again by the similarity with native people who state that it is only on their land that an Inuk is a real Inuk. Anyone who has spent even a brief time in a place like Inuvik, and then travelled with the people on their land, cannot doubt the veracity of that claim.[20]

The questions raised, even by these examples, require careful analysis. We may accept that people should not be casually deprived of their land and their past and turned into a dispossessed urban proletariat. But what *are* their alternatives? What are the consequences of leaving well enough alone? How does a sensitized environmentalist approach these matters, given the fact that ranchers are the ruggedly individualist basis for the exploitation and sterilization of something as close to home as food, and miners the foundation of an endlessly expanding industrialism? Trade-offs between jobs and the environment, between traditional and modern uses of land, are clearly insufficient. It may well be that in tackling these social questions, political ecology will find that they are inseparable from a broader, critical analysis of society itself. Whether that analysis can be both radical and reformist is, perhaps, the most immediate question for political ecology.

As early as 1970 the editors of *Our Generation* were quite specific on the subject. Reformist groups, they charged, 'did not make the connection between violence on the environment and the society that perpetrates that violence.' The ecological crisis was not to be traced to the role of scientists, technology, population growth, or even the 'profit-system,' but rather was 'deeply rooted in the structure of society ... that is based on man's domination and exploitation both of nature and his fellow man ... and is a result of the concentration of power created by the centralisation of energy, material and human resources, and social administration.' On the other side, Brian Johnson, in a 1977 article entitled 'The Primitive Barrier to Political Ecology,' laments the fact that 'what should now be called political ecology has divided the ranks of the environmentally concerned along the lines of political ideology.'[21] One is reminded of the chairman of the Senate committee on poverty when asked some years ago about the defection of several of its key researchers. He is reported to have said that they wanted to change the system whereas the committee was only interested in helping the poor. It may well be that an adequate understanding of what is happening to the natural environment can only proceed from a clear understanding of the key role that political ideology plays.

The connections between ecology and in particular the politics of capitalism – both state and private – are very deep. Daniel Zwerdling has argued that environmentalists should recognize that 'neither pollution nor poverty nor worker insecurity is a separate problem which can be solved on its own. They are three different ways corporations express themselves in maximising profits through the exploitation of people and resources.' The idea that environmental issues are essentially political is taken a step further by André Gorz, for whom ecology is fundamentally anti-capitalist and subversive. In capitalism, he says, 'the only optimum is the accumulation of capital,' whereas ecology is concerned with 'the optimal preservation of resources, of the environment of biological equilibria; it is concerned with the pursuit of maximum durability and use value rather than exchange value; it is concerned with the fullest possible satisfaction and development of human beings, both in their labour and beyond it, rather than with capitalism's maximisation of the efficiency and productivity of labour.'[22]

Marcuse restates the proposition that human well-being is the goal, one to be realized not through increasing consumption and intensified labour, but through liberation from capitalism. The issue, he says, 'is not to beautify the ugliness, conceal the poverty, deodorise the stench, deck the prisons, banks and factories with flowers; the issue is not the purification of the existing society, but its replacement.' Many writers have documented and analysed the ways in which ecological degradation has greater impact upon the poor, both domestically and internationally, than upon the rich. According to Ritchie Lowry, 'some of us pollute a great deal more than others, and this is a function of who has power in society and who does not.' Cy Gonick notes that the ecologists' metaphor of spaceship earth denies 'any distinction between first-class passengers and second-class passengers' and emphasizes that 'some can afford to plan for growth in the "spaceship earth" and actually draw profits from the elimination of the damage that they do, while others certainly can not.'[23]

The poor indeed are in double jeopardy. Pollution effectively redistributes income from poor to rich, for the rich benefit from cheap production and the poor suffer from damnable working conditions and housing invariably located downwind from industry. At the same time, the remedies proposed and implemented for this degradation impose proportionately greater burdens on them as well. The curtailing of energy use by price rationing is a good example, as are increases in downtown parking rates at municipal lots as a means of keeping automobiles out of city centres. The irony, though no irony for them, is that the poor suffer when pollution reduces their house values, and lose again when it is cleaned up and their rents rise.

Politics and ecology are intimately linked in both practice and theory. Di Norcia stated it quite clearly: 'The practices of men toward men and toward nature, and the knowledge of man and of nature must all be transformed together.'[29] This is hardly a novel conception for subsistence societies, but it is one that applies with equal if not greater force to those developed societies that are based, as Morin noted, on three elements: the Cartesian antithesis between self as subject and the external world as object or objects to be manipulated; the idea of 'science, conceived of as an objective discipline which concerns itself neither with its meaning nor its purpose'; and the bourgeois, and later Marxist, view of man as 'the conquistador of nature, who ultimately becomes the Genghis Khan of the solar suburb.'[24]

Morin's criticism of science should properly and particularly include those who claim to be 'social scientists.' The most pervasive danger is that they will either believe that what they preach and practise is a science, or that principles drawn from the natural sciences can comfortably be transferred to the social sphere. This applies not only to behaviourists but also to, for example, 'ecological planners' who would use the principles of ecology to build an environmentalism on the shaky ground of utility. Political ecology offers a means of answering those who would, in attempting to transcend politics in the name of nature and the natural, perhaps unwittingly spread a doctrine that, as Richard Neuhaus observed, 'is strikingly similar to crucial elements of National Socialism.' Such people tend to be part of what William R. Burch calls 'the new puritanism which, in the guise of saving nature, thunders with irrevocable contempt for our species.' In reviewing the Ehrlichs' work, Ansley Coale notes that 'in spite of their proposals ... they are hyperaware of balances and interconnections in ecosystems ... but seem only occasionally aware of balances and interconnections in social systems.'[25] Political ecology should put an end to idle talk about 'people' being the root cause of environmental problems without recognizing, to extend the metaphor, that those steering the spaceship earth dine in first class, not steerage.

Henryk Skolimowski has argued that the major weakness in the ecology movement is the lack of a comprehensive philosophy as its basis, claiming that its attention has been directed almost exclusively 'to the hardware of civilisation, to its economics and technology.'[26] The point applies particularly to political ecology. A philosophy of ecology alone makes little more sense than one of anatomy or biology. While philosophies are inherent in the natural sciences, as many nuclear physicists are well aware, the requirement in the case of political ecology is for a comprehensive and coherent body of philosophy that forms a sounder basis for its practice than empirical instrumentalism. It seems obvious that this can only emerge from a radical perspective in which theory plays a central role. One can ask political ecology, though not ecology *per se*, to develop such a base.

Political ecology stands at the centre of a demand for the incorporation, or reincorporation in new and different ways, of natural science, social science, and the humanities. Science without politics and politics without science are at best intermediate analyses, and poor prescriptive bases for social action. It is a great pity to see such a prominent Canadian scientist as J.C. Ritchie still missing this point entirely. The call is, in a modern guise, much like that of C.P. Snow twenty years ago in his essay on the two cultures.[27] He argued then that while science is good at doing one thing in depth at one time, this leads to weaknesses when trying to assess a much broader range of social and political questions. The arts on the other hand are good at considering many things at once, but rarely go into any aspect in great depth, and show an abysmal ignorance of science and a consequent inability to incorporate its understanding into political decision-making. Science and the arts then, politics and ecology now, must be brought together intellectually for a better understanding of our predicament, and practically in the effort to achieve balance between people and nature.

Both intelligence and industry have brought us to this point; only thought and action together can extricate us. In social relations and in many spheres of modern science we are, as J. Bronowski has observed, not nearly so concerned with describing facts as with creating images.[28] That there is a manifest connection between politics and ecology should be sufficient to propel their joint study forward, and the study of images in language and pictures, in science and politics, may be a good place to begin.

In Claude Jutra's *Dreamspeaker* the child Peter finds release from his physical and psychological bondage in the old Indian and his physical and spiritual world. In liberation, as in prison, there are powerful links between the individual and his human and natural environment; between the land and the water and the people that belong to them. Land, the metaphor for all of his environment, is inseparable from those who people the social environment. In the environmental debate, the idea and the movement of political ecology are founded upon the mutual dependence of these two spheres of existence and the need to explore their relationship as a key to understanding and, if still possible, reversing the directions in which a globalized political economy is leading us. If Jutra is right, the chances are not good.

NOTES

1 See, for example, the Declaration of the European Group for Ecological Action, 12 Dec. 1976; testimony before the Berger and National Energy Board inquiries into the Mackenzie Valley gas pipeline proposals; and testimony before the Ontario Royal Commission on Electric Power Planning.
2 Huxley, *The Politics of Ecology*, Center for the Study of Democratic Institutions (Santa Barbara, Calif. 1963), 1
3 See G. Reichel-Doklmatoff, 'Cosmology as Ecological Analysis: A View from the Rain Forest,' *Ecologist*, 7, no 1 (Jan.–Feb. 1977), 4–11
4 Engels, letter to Schmidt, 27 Oct. 1890; quoted in Vincent Di Norcia, 'From Critical Theory to Critical Ecology,' *Telos*, no 22 (Winter 1974–75), 93
5 *Oxford Universal Dictionary*, 3rd ed. (London 1955), 1537
6 Enzensberger, 'A Critique of Political Ecology,' *New Left Review* (March–April 1974), 3
7 Maddox, *The Doomsday Syndrome* (New York 1972), 161
8 Smith, *The Politics of Conservation* (New York 1966), xi; Bailey, 'Environmentalism, the Left, and the Conserver Society,' *Conserver Society Notes*, 1, no 2 (Summer 1978)
9 J.A. MacDonald, House of Commons Standing Committee on Indian Affairs and Northern Development, *Proceedings*, 22 April 1969, 671; Rodman, 'Four Forms of Ecological Consciousness,' paper presented to the Annual Meeting of the American Political Science Association, Chicago, 3 Sept. 1976, p. 3
10 MacDonald, A.D. Hunt, *Proceedings, ibid.*, 666
11 Rodman, 'Four Forms,' 3; Morin, 'Ecology and Revolution' symposium, *Liberation* (Sept. 1972), 7
12 Di Norcia, 'Critical Theory,' 85
13 T.W. Adorno and M. Horkheimer, *Dialectic of Enlightenment* (New York 1972), 9; quoted in *ibid.*, 88–9
14 *Ibid.*, 90
15 Gruening, 'The Political Ecology of Alaska,' *Science in Alaska, 1951*, Proceedings of the Second Alaska Science Conference, Mt McKinley National Park, 4–8 Sept. 1951; Rogers, *The Future of Alaska* (Baltimore 1962), 56, 59
16 Weisberg, *Beyond Repair* (Boston 1971), 23
17 Friedenberg, review in the *Canadian Forum*, LVI, 667 (Dec. 1976–Jan. 1977), 24; Enzensberger, 'A Critique,' 7; Sills, in *Human Ecology*, III, 1 (1975)
18 See, for example, Weisberg, *Beyond Repair*; James Ridgeway, *The Politics of Ecology* (New York 1971); William R. Burch, *Daydreams and Nightmares: A Sociological Essay on the American Environment* (New York 1971); Sicco Mansholt, 'Ecology and Revolution' symposium, *Liberation*; Murray Bookchin, 'Ecology and Revolutionary Thought,' in his *Post-Scarcity Anarchism* (Berkeley 1971); and Marie Jahoda, 'Postscript on Social Change,' in H.S.D. Cole *et al.*, eds., *Models of Doom: A Critique of 'The Limits to Growth'* (New York 1973)

19 Sills, 'The Environmental Movement,' 26
20 See Hugh Brody, *The People's Land* (Toronto 1975)
21 'The Radical Implications of the Ecology Question,' *Our Generation* (Jan.–Feb. 1970), 3–4; Johnson, 'The Primitive Barrier to Political Ecology,' *Ecologist* (April 1977), 91
22 Zwerdling, 'Poverty and Pollution,' *Progressive*, 37, no 1 (1973), 29; Gorz (under his pen name Michel Bosquet), in 'Ecology and Revolution' symposium, *Liberation*, 5
23 Herbert Marcuse, 'Ecology and Revolution' symposium, 12; Lowry, 'Toward a Radical View of the Ecological Crisis,' *Environmental Affairs*, I, 2 (1971), 351; Gonick, *Inflation or Depression* (Toronto 1975), 290
24 Di Norcia, 'Critical Theory,' 95; Morin, 'Ecology and Revolution' symposium, 7–8; see also William Leiss, *The Domination of Nature* (Boston 1974)
25 Neuhaus, *In Defense of People: Ecology and the Seduction of Radicalism* (New York 1971), 152; Burch, *Daydreams and Nightmares*, ix; Coale, review of P.R. Ehrlich and A.H. Ehrlich, *Population, Resources, Environment*, in *Science*, 170 (1970), 429
26 Skolimowski, 'The Ecology Movement Re-Examined,' *Ecologist* (Oct. 1976), 300
27 Ritchie, 'Northern Fiction–Northern Homage,' *Arctic*, 31, no 2 (June 1977), 69–74; Snow, *The Two Cultures and the Scientific Revolution* (Cambridge 1959)
28 Bronowski, *The Ascent of Man* (New York 1973), 340

3
Economics and the challenge of environmental issues

Peter A. Victor

I stepped out into the night air that nobody had yet found out how to option. But a lot of people were probably trying. They'd get around to it.

Philip Marlowe, in Raymond Chandler's *The Little Sister*

THE NEO-CLASSICAL PERSPECTIVE

Perhaps the most significant event of the nineteenth century was the establishment of the self-regulating market as the predominant institution in the industrial world. For the first time in history, economic activity became identifiable as something distinct from the rest of man's social life. Paralleling this development was the growth of economics as a system of thought designed to comprehend these new arrangements.

Despite the collapse of the self-regulating market, as witnessed in the twentieth century, one lasting result of developments in the nineteenth century is the popular view that there is an important set of issues which are economic rather than political, social, or environmental. Economic growth, inflation, balance of payments, productivity, are some of the 'economic' issues that are commonly given priority over issues that are 'social' and 'environmental.' The choice of priorities is regarded as 'political.' We are assured, for example, that we must have a healthy, growing economy if we are to afford anti-poverty programs, pollution control, and the means to defend our political freedoms.

The fact of the matter is that the modern institutions of business, labour, and government, and the problems to which they give rise, are just as much political and social as they are economic. It is essential that this be recognized in any attempt to understand the present and possible futures of contemporary society, not least with respect to man's interaction with the rest of nature.

One of the most notable and influential attempts to understand the society of his day, and to prescribe for its future, was that of Adam Smith. The year 1976 marked the two hundredth anniversary of the publication of Smith's *Inquiry into the Nature and Causes of the Wealth of Nations*. This was by no means the first treatise on economics, but it was the first to organize the subject matter into categories that modern economists still find extremely familiar. Production, consumption, distribution, trade, prices, employment, and growth each receive due attention within an integrated framework of fact and theory. The comprehensive nature of Smith's work is impressive particularly when compared with the narrow vision exemplified in the typical journal article that today's professional economist is obliged, by his peers, to write.

One result of the undue degree of specialization within economics, as in so many other disciplines, is the neglect of some important aspects of economic and social processes. A few notable exceptions apart, until the late 1960s economists ignored the environmental issues of pollution and resource depletion which are so intimately related to the functioning of the economy. As the decade turned, the 'environment' became the focus of considerable public concern. Since then, a growing number of economists have addressed these environmental issues, principally with the analytical tools and professional biases developed in dealing with the more established parts of economics.

There now exist numerous books and papers by economists on environmental management, pollution as an externality, public 'bads,' input-output and residuals analysis, and on benefit-cost analysis of the use of land, air, water, and mineral resources. It has become common for introductory textbooks on economics to include at least a chapter on the environment and fewer newcomers to economics are being exposed to the myth, so widely taught by economists only a few years ago, that air and water are free because they are abundant.

This increase in the interest of economists in environmental issues has been only a small part of the total academic and governmental response, which has been impressive if only for the time and money it has involved. All sorts of government environmental agencies have been established, new legislation for environmental protection has been enacted, and environmental degree programs have been devised in an attempt to integrate the ideas and information generated by the older and more firmly established academic disciplines.

Now, after a flourish, things have quietened down. The flood of publications on pollution control and resource depletion has moderated and it is an opportune moment for an assessment of how well economists, particularly of the 'neo-classical' variety, have met the challenge of environmental issues.

The argument to be presented in this chapter is that, by and large, economists have failed to come to grips with many of the fundamentals that underlie pollution and resource depletion. Certainly their diagnosis and policy advice have received a cool reception from environmentalists and government alike. Moreover, it will be shown that this failure has implications not only for the contribution of economists to the environmental debate but for the whole fabric of contemporary, 'neo-classical' economics.

The main thrust of this paper is that there are several premises upon which the economic analysis of environmental issues is built that are incompatible with the premises of other participants in the debate, and which preclude economists from perceiving some of the root causes of the problem. As a preface to the main argument, the neo-classical analysis of pollution and resource depletion and the policies that follow from this analysis will be outlined. A critique and suggestions for a new direction follow.

The neo-classical theory of pollution
The neo-classical explanation of environmental pollution is very straightforward. Essentially, it points out that in a market economy based upon the private ownership of the factors or means of production, the self-interested owners of these resources will be induced to do the best for others by doing the best for themselves. Government may be required to ensure that the conditions for competition prevail by anti-trust measures, and it is definitely required to guarantee for all the property rights of each. Given these conditions the neo-classical economy will commit privately owned resources to their most productive uses.

In the conspicuous case of air and to a lesser extent water, the property rights on which the efficiency of a capitalist economy depends generally do not exist. This gives rise to what neo-classical economists refer to as 'externalities.' Although the precise definition of an extenality is still a matter of debate among economists, it may be thought of as an effect of economic activity that lies outside the normal control of market processes. For example, people exposed to the effects of air pollution caused by the industrial production of goods made and sold for profit, have no recourse through the market to obtain financial compensation. This would not be the case if people owned a marketable right to clean air since, under those circumstances, industrialists wishing to pollute the air in their quest for profit would have to buy the right to do so in the same way as they must buy the right to use the other resources that are necessary for production. In the absence of such rights the market cannot properly regulate activities that pollute and there is a *prima facie* case for some form of governmental intervention.

The neo-classical theory of natural resources

The question of property rights and externalities applies equally well to the neo-classical analysis of the depletion of natural resources. Resource economists of this school hold the view that privately owned stocks of resources will automatically be conserved if resource owners foresee future shortages, since it will be profitable for them to restrict the rate of depletion in order to sell later at the expected higher price. According to this analysis the trouble begins when some element of common property exists: for example, if two oil companies drill wells into the same deposit each company has an incentive to withdraw the oil as rapidly as possible for fear that the other is doing just that. Overfishing is merely an extension of these common property circumstances to a renewable resource. Since fishermen do not own the fish until they have caught them, no individual fisherman can sensibly decide to refrain from fishing to allow the fish to grow and to breed. If this view is widely shared by the competing fishermen the very real problem of stock depletion arises.

Neo-classical environmental policies

The environmental policies arising out of the neo-classical analysis of pollution and resource depletion stem directly from the concepts of externality and common property. Externalities can be 'internalized' by the imposition of effluent charges, which many economists argue are more efficient than effluent standards. This approach to pollution control utilizes the incentives that drive the market economy and only indirectly does it affect the property relations in society. However, some economists suggest that these relations should be changed directly to cope with common property situations.[1] For example, to control the effluent discharged into a lake it has been proposed that the regulatory agency should issue saleable rights permitting the owner of the right to discharge a specified quantity of effluent into the lake. Initially a limited quantity of discharge rights would be sold or given away by this agency, thus establishing the maximum amount of effluent that could be discharged into the lake from all sources. Once all the rights were allocated, any company, municipality, or individual wishing to discharge effluent into the lake would be obliged to purchase the necessary rights from those who already owned them. In this way the market would regulate the use of the lake as a recipient of effluent with all the advantages of economic efficiency that are normally claimed for market processes.

The importance of externalities in the neo-classical analysis of environmental issues has raised the question of their magnitude, particularly when environmental policies are being formulated and reviewed. This empirical concern has been met by an increase in the use of benefit-cost analysis for simulating market values for environmental quality. Benefit-cost analysis allows monetary values to be attributed to externalities. For example, the cost of air pollution might be assessed in terms of such items as extra laundry costs, extra house maintenance costs, increased medical bills, and incomes forgone due to illness and premature death. On this basis projects involving an increase in air pollution may be evaluated not only by those costs and benefits that are normally registered in the market but also by accounting for the social costs and benefits caused by externalities associated with such projects.

Material/energy balance
This brief summary represents the core of the neo-classical response to the environmental challenge. All of the ideas embodied in this analysis were, of course, well established in economics long before the upsurge of interest in environmental issues. There has been only one apparently new analytical idea which neo-classical economists have introduced and this is the notion of materials and energy balance to which Kenneth Boulding drew attention in his widely reprinted paper, 'The Economics of the Coming Spaceship Earth.'[2] He pointed out that the economic activities of consumption, production, and trade involve a rearrangement of matter and not a creation of new material. He likened the Earth to a spaceship consisting of a fixed quantity of material subject to a single source of external energy provided by the sun. With this as his perspective, Boulding argued that the Earth's resources should be husbanded with as much care as the supplies of a spaceship.

Since 1966, a number of other economists have taken up Boulding's ideas concerning material flows and material balances and have used them to emphasize the pervasiveness of externalities.[3] Empirical work has begun on the study of the flow of materials and energy through economic processes at all levels and this has highlighted the essential physical links between the depletion of natural resources and pollution.[4]

This new approach to economic activities is particularly insightful for analysing environmental issues. It may be surprising to discover, therefore, that in fact it is not really a new approach at all. Economists as distinct in their orientations as Alfred Marshall and Karl Marx devoted substantial passages in their respective treatises to a description of economic activity in precisely these terms. Marshall, for example, opened Chapter III of his *Principles of Economics* with the statement that: 'Man cannot create material things ... His efforts and sacrifices result in changing the form or arrangement of matter to adapt it better for the satisfaction of his wants ... As his production of material products is really nothing more than a rearrangement of matter which gives it new utilities, so his consumption of them is nothing more than a disarrangement of matter, which diminishes or destroys its utilities.'[5]

Marx, in whose work this theme recurs time and again, cites a statement published in 1773 by the Italian economist, Pietro Verri: 'All the phenomena of the universe, whether produced by the hand of man or by the general laws of physics, are not in fact *newly-created* but result solely from a transformation of existing material. *Composition* and *division* are the only elements, which the human spirit finds again and again when analysing the notion of reproduction; and this is equally the case with the reproduction of value ... and of riches, when earth, air, and water become transformed into corn in the fields, or when through the hand of man the secretions of an insect turn into silk, or certain metal parts are arranged to construct a repeating watch.'[6]

Marshall must be credited with having anticipated an idea which has since attracted the attention of many of his intellectual grandchildren, but at the same time the materials-balance principle is merely an appendage to the main body of neo-classical economics which retains its emphasis on production costs and consumers' preferences. In the contrary case of Marx, the self-described materialist, this view of economic activity permeates his entire theoretical structure.

This is only one of several aspects of Marx's economic structure which differ from that of the neo-classical economists and which are more consistent with the assumptions made in the various non-economic analyses of environmental issues. This matter will be explored further after the following critique of the neo-classical response to the environmental challenge.

GAPS IN THE PARADIGM

This section will be divided into three areas which are by no means
unrelated to each other: (*a*) the theory of value, (*b*) social relations, (*c*)
the state and the economy.

Value
'Value,' says Joan Robinson, 'is just a word.' It is a word that expresses
'one of the great metaphysical ideas in economics.' Just what that idea
is, Robinson is not too sure. She tells us that value 'does not mean
usefulness – the good that goods do us, nor does it mean market prices.'[7]
In her brief historical survey of economists' theories of value she notes
the so-called paradox of value which bothered the classical economists
who had difficulty in explaining why the price of diamonds was so much
higher than the price of water when the latter was clearly of much greater
importance to man. The marginal 'revolution' of the 1870s resolved this
issue by emphasizing scarcities and the significance of the last or marginal unit of a commodity consumed in determining its price.

Robinson also discussed Marx's labour theory of value on which his
analysis of exploitation is founded. Writing in 1937, Maurice Dobb
expressed an important distinction between what he described as these
'two major value-theories which have contested the economic field'
when he said that they have each 'sought to rest their structure on a
quantity which lay outside the system of price-variables, and independent of them; in the one case, an objective element in productive activity
(that is, labour power), in the other case, a subjective factor underlying
consumption and demand (that is, utility).'[8]

Initially utility theory was founded in Bentham's principle of utility by
which is meant 'that principle which approves or disapproves of every
action whatsoever, according to the tendency which it appears to have
to augment or diminish the happiness of the party whose interest is in
question.'[9] Despite the failure of Bentham or anyone else to establish an
empirical measure of utility, economic theorists continued to presume
that the utility which guided man's economic behaviour was something
that, in principle, lent itself to quantification. This notion of 'cardinal
utility' was supplanted by a view of utility as an entity which could not be
added or subtracted but which could be placed in order of greater and
less. Economists determined that this concept of 'ordinal utility' was a
sufficient basis for virtually all of the analytical results that had previously been obtained with cardinal utility. More recently still, economists

have discarded the concept altogether by establishing a pure theory of choice which requires only that the participants in a market have preferences which satisfy a small number of fundamental axioms.[10] It is this theory which lies at the centre of modern welfare economics and which provides the normative foundation for economists' recommendations for public policy, including environmental policy.

Preferences One of the implications of the replacement of cardinal utility by preferences is that economists have virtually precluded themselves from any interest in the reasons people may have for their preferences. At least with cardinal utility people could be said to prefer things for the utility or satisfaction that they expected to derive from their consumption. Obviously, this allows for the possibility that the expected utility on which people base their desires or preferences might be very different from that which they actually do derive in the process of consumption. If this should happen, the achievement of market equilibrium becomes complicated because people will change their preferences as they learn that the commodities they desired and purchased do not give them the utility on which that desire was based. Unless expected and actual utility finally merge for each person in the economy, equilibrium will not be attained and relative prices will continue to change in a way that is inexplicable in the absence of additional assumptions about peoples' learning processes.

Marshall and his successor at Cambridge, Pigou, each recognized this problem and resolved it by identifying utility with both 'the desires which prompt activity and the satisfactions that result from them.' However, the very fact that they admitted this distinction permitted them to consider the dynamic relation between desires and satisfactions such as Marshall did in his brief chapter entitled 'Wants in Relation to Activities.'[11]

The assumption that Marshall and Pigou made – that utility is to be identified with desires and satisfactions – served the same purpose in their economic analysis as the assumption of unchanging preferences does in neo-classical economics. Preferences are fixed points on which the neo-classical system rests and, as such, they are essential for the determination of relative prices. Changing preferences, resulting from the disappointment of consumers who discover in the process of consumption that their preference was misguided, would cause changing relative prices. Since the determination of relative prices is one of the prime objectives of the neo-classical system, it would seem necessary to incorporate a theory of how preferences change into the system. However, this is one area into which economists have seldom ventured.[12]

In summary, the neo-classical theory of value is a theory of relative prices. It is a theory which is built upon a system of individuals' preferences which are assumed, wrongly, to be given and unchanging. In the neo-classical world it is difficult to distinguish value from price. Value is not an explanation of price. It is not even an explanation of preferences. It may be something which people attribute to the things they prefer but it is certainly not a quality that is inherent in the things themselves waiting to be discovered and unlocked in a manner appropriate to the thing itself.

The relevance of all this to environmental issues is that virtually all environmentalists call into question the preferences of people as expressed in the workings of the economy.[13] Far from being content to take them as given, they consider the determination of preferences as an essential aspect of environmental problems.[14] As well, many environmentalists criticize preferences and the actions to which they lead, on grounds that are unacceptable to neo-classical economists. For example, it is suggested that the preferences of individuals and the behaviour based upon them are: (a) not in the best interests of the individuals concerned, and (b) not in the best interest of the environment.

A variety of arguments are used to support these statements and it would be a mistake to think that environmentalists have reached a consensus on these issues. Nevertheless, many environmentalists would subscribe to the following views on preferences, value, and the environment – views which are incompatible with the premises of neo-classical economics.

First of all, environmentalists think that peoples' preferences frequently do not reflect the very real benefits to be obtained from exposure to a healthy environment. In defence of this position, it may be argued that no one would question its validity with regard to children, or those designated as mentally ill, and even a substantial proportion of old people. Those comprising this large section of the population are constantly being told what is good for them by others who claim to know better and, significantly, who are in a position to impose their will. The argument continues that just as children, and other dependents in society, may prefer things which are not in their best interests, so may adults. In support of this position, René Dubos[15] has argued that in the course of man's biological evolution, a number of human requirements were developed. These are not only for nutrition, and protection from the hazards of nature, but also for contact with plants and animals, exposure to the wilderness, and a lifestyle that is consistent with man's hormonal and endocrinal daily, seasonal, and yearly rhythms. His critique of society is based on the observation that people fail to satisfy these requirements because they are too often guided by preferences that bear an imperfect relation to their needs.

The economists' view of value as subjective precludes any discussion of whether individuals' preferences are consistent with their needs. People are *presumed* to be the best judges of what contributes to their welfare. If they make mistakes, then they are to be the judges of that also. In any case, the assumption that preferences are stable through time does not allow for mistakes of a sort which could lead to revisions of those preferences themselves. And relaxing this assumption does not really help since, if it is possible for a person to discover that what he preferred and chose failed to meet with his expectations, then it should be allowed that others could have predicted this occurrence. Such a prediction could very well rest on the observation that what individuals prefer or want does not coincide with their needs.

Information   There is perhaps one avenue open to neo-classical economists whereby a divergence between preferences and needs could be accommodated within an otherwise unaltered framework of economics. The discrepancy between the two could be attributed to a lack of information on the part of the individual.[16] But this creates as many problems as it solves. Is information just another commodity for which an individual may or may not have a preference even though he has a need for it? Is this discrepancy between a preference and a need for information also to be explained by a lack of information? If so, where does the process end? In addition to this difficulty, once economists admit qualifications to the significance they attribute to preferences, other criteria are automatically introduced into the debate. Such criteria, which by definition are external to or additional to the preferences of individuals, transcend the preferences on which the neo-classical system rests.

An environmental ethic   The normative significance of preferences is certainly a point of contention between economists and environmentalists. However, they differ even more fundamentaly over the idea that actions can and should be assessed in terms of their value to the environment. Environmentalists do not agree with economists who insist that the environment has 'value' for individuals only for its utility. They argue that preservation and improvement of the natural environment is important for its own sake and not merely because people wish to make use of 'environmental services.' In other words, they advocate a system of ethics which encompasses animate nature in its entirety and even inanimate nature as well. J. Bruce Falls poses the question: 'And what of the plants and animals themselves? Who gave us the right to eliminate other species? Doesn't our moral responsibility extend to the whole of nature of which we are a part?'[17]

There seems to be no way of *integrating* this view of the environment's intrinsic value into the scheme of neo-classical economics. Once again it calls into question the proper place of individuals' preferences, particularly in the determination of environmental policy. Quite clearly, this is an important area of dispute between environmentalists and economists. At the very least, its resolution will require on the part of economists a recognition of the difference between preferences and needs, and an investigation into the empirical relationship between preferences, needs, and welfare within a specific social context. In the meantime, environmentalists will continue to resist the lessons of a system of economics which presumes that individuals' preferences are the only legitimate criterion for establishing and assessing environmental policy.

The social relation
The basic unit in modern economics is the individual, whereas in sociology it is the social relation.[18] The only social relation, that is, relation between people, that is fully recognized in economic theory is the market relation between self-interested buyers and sellers. Some economists have examined various implications of interdependence among peoples' 'utility functions,' but significantly all the theorems of welfare economics concerning the efficiency of markets are based on the assumption that these utility functions are not interdependent. Indeed, these theorems *require* the assumption that there are no other social relations except market relations. This is immediately relevant to environmental issues in two ways:
1 / economists frequently attribute excessive pollution and resource depletion to market failure rather than to the breakdown of other forms of social relations;
2 / the environmental policy recommendations of economists are usually recommendations either to extend the scope of the market where it presently fails to operate, for example, by establishing pollution rights or

effluent charges, or to simulate the market by benefit-cost analysis so that government can regulate what the market would declare if circumstances allowed it to function properly.

In neglecting other forms of social interaction, economists account for environmental problems in terms of the failure of market relations and recommend that the situation be remedied by a direct or indirect extension of the market for the regulation of environmental 'resources.' In contrast, environmentalists, to varying degrees, attribute this type of problem not to the limited scope of the market, but to its over-pervasiveness in modern social life. They express the concern that market decisions may be too short-sighted or that the future will be too heavily discounted, or that advertising, which is an integral part of the market, encourages too much consumption and production of the wrong type of goods and services. Policy recommendations which emanate from this line of analysis usually call for greater restrictions on the market, say in the form of legislated regulations rather than the extension of the market, which economists are predisposed to favour. Such a line of thought also makes environmentalists tend to reject the assumption that all things of 'value' are or should be for sale. For environmentalists the relevant question is often: 'Should a person have the right to pollute?' Whereas an economist would ask: 'If the right to pollute were saleable, what price would it fetch?'

The market and the dissolution of non-market social relations   There is an interesting and dynamic side to the question of market and other types of social relations. Numerous examples exist of over-exploitation of the environment because of the collapse of social restraints based on tradition, myth, and custom following the geographic expansion of capitalist economies. Richard Cooley's study of the Alaskan salmon fishery shows how property rights and the ownership of a fishing site were held by specific tribes and clans.[19] The salmon were a group totem for the Alaskan Indians who identified their genealogical continuity with the migrating cycles of the salmon and so were particularly careful not to deplete the fish stock. Over-exploitation of the fishery began only when the Indians came into contact with the rest of North America and the fish became a marketable commodity. The cultural checks that historically had preserved the fishery were destroyed, giving rise to an example of what would currently be termed the 'tragedy of the commons.'[20] The same tragedy, caused by a complex of factors, afflicts our east coast fisheries.

The experience of the North American Indian clearly illustrates the way in which the spread of the market undermined traditional relationships between man and man and, hence, between man and environment.[21] S.L. Udall observed that 'The land and the Indian were bound together by the ties of kinship and nature, rather than by an understanding of property ownership ... the Indian's title, based on the idea that he belonged to the land and was its son, was a charter to its use – to use in common with his clan or fellow tribesman, and not to *use up*.' With the growth of the fur trade in North America, E.B. Leacock noted, 'The new trade introduced new bases for both cohesion and fragmentation in Indian society. Trade for basic necessities loosened the economic basis for cohesion upon which the "stateless society" was based. The introduction of commodity exchange cut at the reciprocity of basic relations in Indian society and weakened the foundation for traditional forms of leadership.' H. Hickerson, who has investigated this process in considerable detail as it applied to the Chippewa of the Upper Great Lakes, emphasized three factors: '(1) changes in social norms ... related directly to the imposition of outside socio-economic and ideological systems that everywhere gathered momentum; (2) such outside systems were ... exploitative and continue to be so; and (3) changes in organization ... in the long run did not result in accommodation to real relations, that is, relations of exploitation.' He goes on to say that 'One factor seems ... of overriding importance and must apply at one time or another everywhere trade was instituted. This was a factor growing out of the very conditions of trade, at the same time generating its social relations. This is the factor of the *individualization of the distribution of food*.'[22]

The North American experience of the spread of the market and its destructive influence on the rights and duties which had traditionally regulated man's relations with nature is useful because it is relatively easy to see how the process operated as European capitalism moved across the American continent. A less obvious but equally important example of a similar process is found in the enclosure of agricultural land in pre-industrial England. Prior to enclosure, which was based on the desire of the more prosperous farmers to produce for the growing city markets, the villagers farmed open fields. Each had access, restricted by kinship and village custom, to the common land for pasturing animals and obtaining firewood and building materials. When enclosure included the common land, it reduced the marginal cottagers and small-holders, who already sold any surplus produce on the market, to 'simple wage labour. More than this: it transformed them and the labourers from upright members of a community, with a distinct set of *rights*, into inferiors dependent on the rich.'[23]

Karl Polanyi, more than anyone else, has stressed the historical significance of the development and ultimate collapse of the self-regulating market. He made the following observations about the establishment of markets in labour and land in Europe during the eighteenth and nineteenth centuries: 'To separate labour from other activities of life and to subject it to the laws of the market was to annihilate all organic forms of existence and to replace them by a different type of organisation, an atomistic and individualistic one.' And about land he said: 'What we call land is an element of nature inextricably interwoven with man's institutions. To isolate it and form a market out of it was perhaps the weirdest of all undertakings of our ancestors.'[24]

Polanyi's observations suggest that the market has been destructive of the various forms of non-market controls which served to regulate man's relation to man and to the environment in such a way that avoided the over-exploitation of communal land. Thus while the economist's analysis of the despoliation of common property resources has an undeniable relevance to today's world, it does not follow that, since the market and common property are incompatible, it is common property which has to go. These examples from history indicate that the expansion of the market system *created* the tragedy of the commons by weakening the traditional forms of social relations which had hitherto prevailed. It is the reconstruction of social structures such as these, combined with the propagation of an environmental ethic, that environmentalists argue is an essential ingredient in an effective environmental policy. Since the market tends to weaken and destroy such structures by undermining all social relations except those of purchase and sale, environmentalists look to control of the market rather than to market control as the proper framework for resolving environmental issues.

The state and the economy

Neo-classical economic analysis is primarily concerned with the functioning of a market economy based on the private ownership of property. Its analysis shows that in the absence of certain conditions relating to the distribution of property and the level of expenditure in the economy, in addition to those conditions necessary for perfect competition, a capitalist economy will be inequitable, inefficient, and subject to unemployment and inflation. If these defects are to be remedied, it is the task of the public sector as prescribed in the theory of public finance.[25]

This theory of public finance, which has its origins in Smith's *Wealth of Nations*, is exclusively normative. It is a theory of what the state ought to do so as to improve the functioning of the economy. It is not a theory to explain the role of the state in capitalist societies, which is thought by neo-classical economists to be the proper concern of political scientists. In a structural sense, therefore, the state is exogenous to the neo-classical framework. Consequently, economists who analyse environmental problems in terms of market failures of one kind or another resort very readily to the recommendation of solutions to be imposed by the state.

In terms of practical politics, most environmentalists also look to the state to resolve the environmental issues with which they are concerned. However, they do this with varying degrees of scepticism about the impartiality of the state. Some adopt the position that the state is so intimately involved with the interests of private property that it cannot be relied upon to protect the environment if such protection is incompatible with the interests of private property.

Views such as this are derived from some notion of what the state is and how it functions in capitalist societies. These are issues to which economists of the neo-classical school have not traditionally addressed themselves. And yet, in taking for granted, as they do, that the state is exogenous to the market, these economists base their analysis of problems and recommendations for their solution on the underlying assumption that the state is independent and impartial – an assumption which is at variance with the views and the experience of many people. If economists are to improve their understanding of economic and environmental issues in this era of big government, it is necessary for them to incorporate into their analysis a theory of the state which goes well beyond the traditional one of public finance and the unduly simplistic assumptions of neo-classical economics.[26]

OLD ORIGINS, NEW DIRECTIONS

Whenever the mainstream of economic thought encounters difficulties, there is an opportunity for the ever-ready Marxist critique of 'bourgeois' economics to surface. A case in point was *Business Week*'s publication of a review of Marxist analyses of inflation and depression in 1974, when it was clear that more conventional economists were floundering. It is significant, therefore, that there has been very little Marxist economic analysis of environmental problems.[27] The simplistic response typified by a leading Marxist economist, Ernest Mandel, is clearly inadequate: 'The evil is private property and competition, that is, the market economy and capitalism. All catastrophes, including the irrational and inhuman roads that technology is led down, derive from this social base and from it alone.'[28]

This particular form of reductionism is definitely a misrepresentation of those ideas of Marx which can be related quite easily to the environmental issues of today. It has already been noted that Marx gave considerable prominence to the materials-balance principle which has attracted the attention of many economists in recent years. Moreover, his broadly conceived analytical framework is not open to the same criticism about social relations and the state that can be levelled at the neo-classical framework. This section will highlight some of the major parts of Marx's analysis which are of particular importance in understanding environmental issues and which can provide a new direction for economics of the environment.

Marx's view of man and nature

As an indication of the relevance of some of Marx's ideas to environmental issues, consider his treatment of man's relation to nature. For Marx, this relationship is dialectical. Those who have wrestled with the problem of whether man is a part of nature or is apart from nature find it refreshing to discover Marx's view. He says that 'Man opposes himself to Nature as one of her own forces ... in order to appropriate Nature's production in a form adapted to his own wants. By thus acting on the external world and changing it, he at the same time changes his own nature.'[29]

There are two important ideas contained in this quotation: (1) man is regarded as a part of nature, albeit, that part of nature which is self-conscious; and (2) man changes the world by his labour and, at the same time, changes himself. Neither of these ideas are found or can be accommodated in the neo-classical framework. There the distinction between man and nature is so sharply made that it is customary for neo-classical economists to define economics in terms of human wants on the one hand and scarce resources on the other.[30] These wants or preferences are assumed to be immutable and usually insatiable, and nature is regarded as no more than a set of resources for satisfying these wants. In Marx's analysis, needs arising from people's natural requirements, which are more or less fixed, are distinguished from needs which have a social origin and are therefore subject to change. This distinction is consistent with the view of human needs to which environmentalists such as René Dubos subscribe.[31] Moreover, Marx is able to integrate it into his framework without the difficulties that changing needs and preferences involve for the neo-classical school. This is so because the cornerstone of Marx's economic analysis is not an abstract notion of need, utility, or preference, but labour, which for Marx represents a direct empirical interaction between man and nature.

The neo-classical view of nature as a set of resources for satisfying human wants is not only recognized by Marx as a view that is characteristic of the age of capitalism, but he seeks to explain it in those terms. Referring to the advent of capitalism, Marx says: 'Nature becomes for the first time simply an object for mankind, purely a matter of utility; it ceases to be recognized as a power in its own right; and the theoretical knowledge of its independent laws appears only as a strategem designed to subdue it to human requirements. Pursuing this tendency, capital has pushed beyond national boundaries and prejudices, beyond the deification of nature and the inherited self-sufficient satisfaction of existing needs confined within well-defined bounds, and the reproduction of the traditional way of life. It is destructive of all this, and permanently revolutionary, tearing down all obstacles that impede the development of productive forces, the expansion of needs, the diversity of production, and the exploitation and exchange of natural and intellectual forces.'[32]

That quotation brings out another characteristic of Marx's analysis which sharply distinguishes it from neo-classical economics. His work is infused with a sense of history. He was concerned with uncovering the 'laws of motion' of capitalist society and he emphasized the origins of capitalism as well as its basic instability and impermanence, which he claimed to have demonstrated. By analysing capitalism in its historical context, Marx discerned that capitalism transforms everything of value into commodities to be bought and sold, and this includes people as labour power and nature as resources. In contrast, the practice of neo-classical economists is to use the historically specific phenomenon of the all-pervasive market for analysing contemporary environmental issues and to search for a solution within the same restricted historical setting. Many environmentalists, for their part, do not presume that solutions are possible without far reaching socio-economic changes,[33] and this provides them with a common perspective with Marx on the *need* for societal change if not on its *inevitability*.[34]

Among Marx's other observations which are of direct relevance to environmental issues is the attention he devoted to the various forms of property that have occurred in history. He commented on the durable, common property arrangements amongst 'Romans, Teutons, and Celts as well as Indians and Slavs.' He also noted the environmental implications of the increasing urbanization of the population: 'Capitalist production, by collecting the population in great centres, and causing an ever increasing preponderance of town population, disturbs the metabolism of man and the earth, i.e. the return to the soil of its elements consumed by man in the form of food and clothing, and therefore, violates the eternal condition for the lasting fertility of the soil.'[35] This theme, which has since been taken up by Barry Commoner in *The Closing Circle*, was presented by Marx together with another pertinent observation of his. Having argued that capitalism transforms everything of value into commodities, including labour and nature, Marx goes on to say how both of these are exploited by the same process: 'All progress in capitalistic agriculture is a progress in the art, not only of robbing the labourer but of robbing the soil; all progress in increasing fertility of the soil for a given time is a progress toward ruining the lasting sources of that fertility ... Capitalist production, therefore, develops technology, and the combining together of various processes into a social whole, only by sapping the original sources of all wealth: soil and labourer.'[36]

The importance of production

Apart from its obvious and explicit relevance to the economics of environmental issues, this last idea reflects what may be the single most significant feature of Marx's analysis for reformulating the economics of environmental issues, the emphasis on the process of production: 'It is a distinctive feature of the Marxist approach to economics – in common with the classical economists – that the central focus of its analysis is productive relations ...'[37] In recent years there has been a renewed effort by some economists to reinstate the analysis of production as the central theme in economics. The most significant analytical work has been that of P. Sraffa, whose economic analysis has been complemented at the philosophical level by the work of M. Hollis and E.J. Nell, who call for the abandonment of positivist, neo-classical economics and its replacement by a rationalist, classical-Marxian framework.[38]

The thrust of this Marxian approach is to establish the logical implications of a system of production comprising numerous interdependent productive activities. To provide a viable basis for any society, a system of production must be capable of reproducing itself. This means, for example, that the production of each commodity in each year must be sufficient to meet the needs of the economy during the year and to replenish the initial stocks that existed at the start of the year. This is equivalent to Marx's schema of 'simple reproduction.'

It is easy to see the relevance of examining the conditions for a reproductive economic system in terms of environmental issues. One of the threats of resource depletion and environmental pollution is that an economic system which gives rise to these problems may be incapable of resolving them. This raises questions about the sources and nature of technological change and whether such change alleviates or aggravates the environmental obstacles to an economic system's capacity to reproduce itself. Technological change is part of the larger social process by which people respond to the situations with which they are confronted. One of the contributions by Forrester and Meadows[39] to the debate about the limits to growth was their success in drawing attention to the existence of positive and negative feedbacks within the socio-economic system which can either dampen or aggravate explosive trends. For example, if a growing city responds to traffic congestion by increasing the capacity of its transportation systems, then this will tend to increase the growth of the city and ultimately worsen the transportation problem.

Whereas Forrester and Meadows were primarily concerned with the behaviour of a given system, albeit the world system, their perspective is of even greater significance when we consider the evolution of interacting sub-systems within society. Economic growth in the modern era has required and encouraged an increasing mobility of labour. Despite improvements in communications, this has undermined the viability of the extended family, which used to perform many of the welfare functions that are now performed by the state. It is an open question whether or not the nuclear family is a self-sustaining social unit, and the experimentation in the past decade with various forms of communal life suggests that some people at least are trying to create circumstances in which they can have closer contact with a larger group of people than is possible within the nuclear family.

The lesson to be learned here is that, even if sustained economic growth is capable of continuing indefinitely via the successful substitution of new resources for old, it is unlikely to bring the social, political, and environmental stability that the advocates of growth argue can only be bought with increased wealth. Instead, the process of economic growth, with its requirements for narrowly defined economic efficiency, may preclude the decentralization of economic and political activity, towards which an increasing number of people are looking as a condition for coping with many of our contemporary social and environmental problems.[40] In other words, not only must the economic system be capable of reproducing itself, but it must do so in a way that is consistent with reasonably stable social systems. Whether or not this will be possible without once again submerging man's economy into his social relations remains to be seen.[41]

Steady-state economics

It is ironical that Kenneth Boulding, who has spoken out publicly and indiscriminately against the economic theories of Marx, should have written the seminal paper for environmentalists, on what has come to be known as steady-state economics. He introduced to many the idea of materials throughput, which, as shown above, is entirely consistent with the Marxian perspective. Boulding questioned the viability of any economy as dependent as ours on using up the non-renewable resources provided by nature and he advocated a substantial shift to the use of renewable resources, particularly solar energy.[42]

More than a century before Boulding, the influential classical economist, John Stuart Mill, regarded the attainment of the stationary state not only as inevitable but as definitely desirable. He noted that 'it must always have been seen, more or less distinctly, by political economists, that the increase in wealth is not boundless: that at the end of what they term the progressive state lies the stationary state ... [which] I am inclined to believe ... would be, on the whole, a very considerable improvement on our present condition.'[43] How remote this vision from the current musings of neo-classical economists!

CONCLUSION: FROM THEORY TO POLICY[44]

The increasing pervasiveness of the market in human social life, which has had serious implications for people's interactions with one another and with the natural environment, has been the result of a complicated historical process rather than a conscious social decision. Neo-classical economists have provided striking insights into the market system without considering the means by which it became established. For many economists, not least those concerned with environmental issues, advocacy has followed analysis. Consequently these economists have helped carry forward, albeit in a small way, a historical process of which they are scarcely aware.

Despite the vigour with which economists have proposed market-based solutions to environmental problems, there is yet to be a deliberate and widespread extension of the market, and the mentality to which it gives rise, to regulate the use of the full range of 'environmental resources.' One reason for this may be that bureaucratic relations have begun to supersede market relations and those with bureaucratic power are reluctant to relinquish their decision-making authority to market processes which they do not fully understand and which they are less able to control.

It is of current concern, therefore, whether those aspects of man-environment relations which are outside the market at present should remain there or whether they should be brought directly or indirectly under market control. Decisions relating to the environment can be social and political rather than economic. Economic considerations can still play a part, but the institutional processes by which environmental decisions are made can allow for the generation and flow of information and a distribution of influence, power, and responsibility that differs significantly from that which prevails in the market.

In the past the extension of the market has been thrust upon people, sometimes in the face of their opposition and frequently with the support of the state. At this point in history an opportunity exists for making a conscious decision about the exclusion of the environment from the market system. A decision to continue its exclusion and even to extend it, so that, for example, the exploitation of natural resources might become subject to publicly determined quotas, presents opportunities for restructuring our social and political decision-making institutions.[45] Some attempts to do this in Canada are already in evidence with the establishment of some form of environmental assessment in each province and by the federal government, and the somewhat novel approaches to their tasks of such Commissions as the Berger Commission and the Royal Commission on Electric Power Planning. It is still to be seen whether these initiatives will serve only to bolster the existing institutional structures or whether they are capable of leading the way to institutional changes adequate to cope with the environmental problems with which we are confronted.

NOTES

1 J. Dales, *Pollution, Property & Prices* (Toronto 1968)
2 Boulding, in H. Jarrett, ed., *Environmental Quality in a Growing Economy* (Baltimore 1966)

3 R.U. Ayres and A.V. Kneese, 'Production, Consumption and Externalities,' *American Economic Review*, LIX (June 1969), 282–97. See also N. Georgescu-Roegen, *The Entropy Law and the Economic Processs* (Cambridge, Mass. 1972). This book is especially noteworthy in that it represents an attempt to reconcile economic activity with physical laws of the twentieth rather than the nineteenth century.

4 P. Victor, *Pollution: Economy and Environment* (Toronto 1972)

5 Marshall, *Principles of Economics*, 8th ed. (London 1920), 53, 54

6 Marx, *Capital*, I (London 1970), 43

7 Robinson, *Economic Philosophy* (Middlesex 1965), 47, 29

8 Dobb, *Political Economy of Capitalism* (London 1937), 12

9 J. Bentham, *Introduction to the Principles of Morals and Legislation* (London 1948), 1

10 H.A.J. Green, *Consumer Theory* (Middlesex 1971), 22–5

11 Marshall, *Principles of Economics*, 78n, 73–7

12 Cf. Green, *Consumer Theory*, 26: 'Such considerations are peripheral to standard textbook discussions of consumer behaviour and for the most part remain so in this one.'

13 See, for example, W. Leiss, *The Limits to Satisfaction* (Toronto 1976)

14 Some economists, notably J.K. Galbraith, have addressed this issue but their work lies outside the mainstream of contemporary economics.

15 Dubos, *A God Within* (New York 1972)

16 As implied by Green's comments on advertising, information, and preferences. *Consumer Theory*, 27

17 Falls, 'The Importance of Nature Reserves,' in B.M. Littlesohn and D.M. Pimlott, eds., *Why Wilderness?* (Toronto 1971), 26

18 Marx's emphasis on social relations accounts for the much greater attention he receives from sociologists than from most contemporary economists.

19 Cooley, *Politics and Conservation* (New York 1963)

20 G. Hardin, 'The Tragedy of the Commons,' *Science*, 162 (13 Dec. 1968), 1243–8

21 An excellent account of this process is given by Irene Spry in 'The Great Transformation: The Disappearance of the Commons in Western Canada,' *Man and Nature on the Prairies*, Canadian Plains Studies 6 (Regina 1976)

22 Udall, *The Quiet Crisis* (New York 1963), 5–7 (emphasis in original); Leacock, *Introduction to North American Indians in Historical Perspective*, ed. Leacock and N.D. Lane (New York 1971); Hickerson, 'The Chippewa of the Upper Great Lakes: A Study in Sociopolitical Change,' in *ibid.*, 183, 186 (emphasis in original)

23 E.J. Hobsbawm, *Industry and Empire* (Middlesex 1968), 102 (emphasis in original)

24 Polanyi, *The Great Transformation* (Boston 1944), 163, 178

25 R.A. Musgrave, *The Theory of Public Finance* (New York 1959)

26 See, for example, R. Miliband, *The State in Capitalist Society* (London 1969)

27 Two items of note are H. Rothman, *Murderous Providence* (London 1972), and R. England and B. Bluestone, 'Ecology and Class Conflict,' in H. Daly, ed., *Toward a Steady-State Economy* (San Francisco 1973)

28 Mandel, 'The Generalized Recession of the International Capitalist Economy,' *Inprecor*, 16 Jan. 1975, p. 16

29 Marx, *Capital*, I, 177

30 L. Robbins, *An Essay on the Nature and Significance of Economic Science* (London 1937)

31 See pages 42–3 above

32 Marx, *Grundrisse* (Middlesex 1974)
33 See, for example, 'The Blueprint for Survival,' *The Ecologist* (Jan. 1972). This document represents a modern example of what Marx referred to, somewhat disparagingly, as utopian socialism. Its omission of any mention of private property is striking.
34 Exponents of the 'Limits to Growth' thesis do concur with Marx on the inevitability of collapse if certain harsh measures are not quickly introduced. They are notably less optimistic than Marx about the outcome of such a collapse should it occur.
35 Marx, *Capital*, I, 505
36 Commoner, *The Closing Circle* (London 1971); Marx, *ibid*. It should be noted that the somewhat favourable interpretation given here of the relevance of Marx's work to environmental analysis is at variance with Georgescu-Roegen's view that 'both main streams of economic thought view the economic process as a "no deposit, no return" affair in relation to nature.'
37 J. Eaton, *Political Economy* (New York 1966), 7
38 Sraffa, *Production of Commodities by Means of Commodities* (London 1960); Hollis and Nell, *Rational Economic Man* (London 1975)
39 D. Meadows *et al.*, *The Limits to Growth* (London 1972)
40 E.F. Schumacher, *Small Is Beautiful* (New York 1973)
41 K. Polanyi, 'Our Obsolete Market Mentality,' reprinted in G. Dalton, ed., *Primitive, Archaic and Modern Economies* (Boston 1968)
42 Public lecture at the University of York, and 'The Economics of the Coming Spaceship Earth'
43 Mill, *Principles of Political Economy*, II (London 1857), 320
44 I am grateful to Abraham Rotstein for suggesting the logic of this conclusion.
45 See Daly, *Toward a Steady-State Economy*

4
The Toronto lead-smelter controversy

C.C. Lax

So far as attempts to evaluate the environmental impact or consequences of certain activities are concerned, it is increasingly obvious
that 'environmental issues' have been misnamed. In light of my experience in the lead controversy in Metropolitan Toronto, I have concluded
that in every so-called 'environmental issue' the ecological factors have
been subrogated to the political, social, and economic factors that are
inevitably present. Indeed, as a lawyer, I have just begun to understand
fully the concept that pollution is only the end result of the society in
which we live and that it is our society, rather than certain isolated
activities, which gives rise to the severe environmental contamination
which has caused such great concern.

There has been considerable reluctance on the part of the government
officials to act effectively to prevent environmental degradation or to
protect public health from dangers caused by harmful substances secre-
ted into the natural environment. Perhaps they appreciate and fear the
inevitable questioning of our social and political systems that must form
a part of the solution. Indeed, the usual response is to trade off the
environmental concerns in order to protect the economic system and our
present social make-up.

This conclusion is justified when viewed in the light of governmental
response to environmental hazards created by the presence of lead,
mercury, arsenic, polychlorinated biphenyls (PCBs), asbestos, and, most
recently, radiation. Government action is directed solely to the eradica-
tion of the traces of contamination, if possible, without serious examination of the source or cause. In this process, the government must
naturally, if somewhat illogically, attempt to balance the need for
economic growth against environmental interests. Ultimately, the decision is based upon political considerations and, unless politically expe-
dient, short shrift is given to the environmental interest. The short-term
politically attractive solution is preferred to the long-term and ultimately

more sensible approach. The over-all goal is to maintain the economic, social, and political *status quo* while placating the environmental interest groups.

That the process is illogical is apparent. That it is dangerous is for future generations to bear out. That it is unjust and discriminatory against those least able to fight back is, I hope, to be demonstrated by the following discussion.

The text that follows is a discussion of lead contamination in the Metropolitan Toronto area. The problems and frustrations I encountered in dealing effectively with the issue of lead contamination make it impossible to ensure any dispassionate objectivity.

LEAD CONTAMINATION IN THE METROPOLITAN TORONTO AREA

Lead is a toxic substance which has been recognized as a danger to human health from the earliest times. It is somewhat surprising, therefore, that people are still exposed to lead in many ways, with the most common sources being paint, gasoline, and industrial processes. In the recent Toronto experience, the public exposure to lead was caused mainly by secondary lead refiners and smelters. In their efforts to defend themselves from being identified as the culprits, the industries involved attempted to deflect public attention to the other possible sources.

In view of the widespread and ancient knowledge of the harm that lead could inflict upon health, it is somewhat shocking that it was not until 1973 that our provincial officials began seriously to analyse the health hazard in lead-contaminated areas. There is evidence to suggest that the public, especially young children, were exposed to serious levels of lead contamination in parts of Toronto as early as 1965. In that year five children living in the vicinity of Toronto Refiners and Smelters Limited were treated for lead intoxication at the Hospital for Sick Children. Health authorities had no difficult search to locate the possible stationary sources. Besides Toronto Refiners, there are only two other secondary smelters or refiners in Toronto – the Canada Metal Company Limited and Eltra of Canada Limited (Prestolite Battery Division) – and two in Mississauga – ESB Canada Limited and Tonolli Company of Canada Limited. All five were known to both the Ministry of the Environment and the Ministry of Health as potential sources of lead contamination.

The issue of lead contamination first became public in April 1970, when a citizens' petition was presented to the Ontario government, seeking its assistance in curbing the dust emissions from the premises of Kaufman Metal, a predecessor of Toronto Refiners and Smelters. A lead emission analysis was started in that area in September 1970, which apparently produced results that failed to arouse governmental suspicion. The government investigators at that time failed to take notice of a battery-crushing operation that was carried on in the open and for which the required environmental approvals had never been granted.

In 1972 the citizens, tired of government inaction, sought the aid of the Canadian Environmental Law Association. With their assistance, public pressure was brought to bear on the Ministry of the Environment, who were prevailed upon to shut down temporarily the offending operation. From 1970 to 1972, dust samples had been collected by the affected citizens themselves, all the time quite unaware of the high lead levels those samples would contain.

Once the Ontario Ministry of the Environment became aware of the potentially hazardous situation around Toronto Refiners and Smelters in 1972, they did not display any appreciable concern for the health of surrounding residents. They failed to advise the public of the potential health hazard, failed to set up blood-testing facilities for people living in grossly contaminated areas, and failed to insist upon strict adherence to industrial emission abatement programs in other lead-processing plants. For over a year the government officials observed industrial practices which contravened not only existing legislation but also basic principles of industrial health.

Finally, in January 1973 a citizens' deputation enlisted the help of the Local Board of Health of the City of Toronto. For the first time the citizens were given a sympathetic hearing, with the promise of some assistance from elected officials. What followed was an increasingly bitter confrontation between municipal and provincial officials. During the course of the confrontation it was discovered that a potentially more serious problem existed in the east end of the city, around the premises of the Canada Metal Company Limited.

With the entry of Canada Metal into the picture, the debate shifted focus. We were no longer concerned with the relatively small-scale activity of a converted junk yard (Toronto Refiners), but with the industrial activity of a large and relatively modern lead refiner and manufacturer. Unlike Toronto Refiners, which was privately owned, Canada Metal was controlled in part by Cominco, the mining arm of Canadian Pacific Investments, and in part by National Lead, a us lead producer and refiner.

The two lead smelters joined together in a common front against the mounting pressure. They retained the same law firm, public relations firm, and consulting engineers to assist them. There was the clear concern that, if they were forced to adopt radical and expensive pollution abatement measures in Toronto, their industry would be faced with similar demands in other parts of Canada and the United States.

As the affected citizens became more vocal, the suspected sources of contamination became more active in their denial of liability and denunciation of the citizens' claims. Press releases were issued flatly denying any contribution by these refiners to the levels of air-borne lead. Experts were retained and brought to Toronto to testify before the Local Board of Health, stating that air-borne lead posed no health hazard and that we need only be concerned with lead that was ingested rather than inhaled. Lurking in the background were the officials of the industry association, the International Lead Zinc Research Organization (ILZRO). This trade spokesman had, for a number of years, developed its own expertise in public health matters related to lead contamination. It maintained an expensive coterie of prominent scientists whose research was in whole or in part funded by ILZRO. These scientists were sent to testify in Toronto on numerous occasions.

The emotional aspects of the issue were further intensified as a number of children living in the affected areas were admitted to the Hospital for Sick Children in Toronto for chelation treatment to reduce high levels of lead in their blood. The media publicity surrounding this treatment only served to heighten public concern about the lack of effective action by the government. Citizens were frustrated by the response of the provincial authorities who in turn accused the citizens of gross exaggeration and panic. The government's appeal for calm was accompanied by numerous statements indicating that the problems would be quickly rectified through the continued co-operation of government and industry.

The citizens in the areas surrounding both Toronto Refiners and Canada Metal organized themselves into residents' associations, called the South of King Street Residents' Association and the BREMM Association respectively. Retaining legal counsel, they maintained a high degree of pressure on the Ministry of the Environment. By the latter part of 1973 they were further assisted by the technical expertise of members of the Institute of Environmental Studies at the University of Toronto. These scientists, using elaborate testing procedures, concluded that the suspected points of emission, such as smokestacks, were probably not the culprits. In their view, the emissions originated from fugitive sources or sources incapable of definite and clear description.

The initial reports of the institute suggesting that these fugitive emissions were the main cause of the problem were met with scorn and sarcasm by the Ministry of the Environment. The ministry held to the view that the problem was caused by stacks and that new and higher stacks together with scrubbing devices would solve the problems.

Valuable time and expertise were wasted in this scientific squabble, pitting the government against the university's scientists. The government's hard-line position accusing the citizens and the University of Toronto of gross exaggeration was maintained until 1974. The citizens were at first shocked and then angered. From the outset they had believed that their concerns were not of particular importance to Queen's Park, but the callous repudiation of their allegations encouraged their resentment.

It was interesting to observe the not so subtle changes in their attitude. Initially, they had believed that the problem would be quickly rectified once the government authorities had had an opportunity to look into their allegations, but as time passed and as their concerns were offhandedly dismissed, they lost confidence in the agencies they had believed would help them. As they saw the government initially align itself with industry, they realized that they were the potential victims in a political battle. This realization led to intense bitterness.

The citizens wondered aloud whether the government response would have been the same if other wealthier areas of the city had been affected. On more than one occasion, officials of the Ministry of the Environment were asked if they would tolerate a similar level of contamination in Rosedale or Forest Hill. This resentment coupled with the lack of response from the elected members of the legislature drove home the realization that they were powerless to fight back unless they took further steps to publicize their plight.

As the public furore increased, the industries decided to fight back using every possible avenue of attack to attempt to stifle the debate. Basically, the brunt of the attack was borne by members of the news media who were subjected to libel and slander proceedings, injunction proceedings, and, later, contempt of court proceedings for their failure properly to obey the court injunctions. These court proceedings against members of the media succeeded to some extent in diffusing the potentially explosive situation and the issue of lead contamination did not receive the same full coverage in the media that it had previously enjoyed.

After the initial offensive against the media the smelters shifted their attention and began a legal attack against the Local Board of Health of the City of Toronto. An allegation of bias was levelled at three members of the board and an order was sought enjoining those members and the board from dealing with the issue of lead contamination. The evidence in support of the allegation of bias was based on the media reports of what the individual members of the board had allegedly said about the industries involved. All of the reporters were called as witnesses. This case was subsequently withdrawn by the smelters in the midst of submission of argument to the Supreme Court of Ontario.

As well as going after public figures, officials, and the news media, the solicitors for the smelters attempted to silence the strident voice of a small left-wing radical group, the Canadian Liberation Movement. This attempt ended in failure and with some embarrassment to the refiners since the president of that group used the opportunity provided by his court battle not only to acquit himself but to publicize further the unwarranted efforts of the refiners to stifle public debate over the issue of lead contamination.

Finally, in October 1973, following the release of preliminary results of blood tests, the Ministry of the Environment issued its first stop order against Canada Metal. This order was hurriedly issued in view of the alarming results of the blood tests. Immediately, the company challenged the validity of the order and it was subsequently set aside by the Supreme Court of Ontario on the basis that it had been an arbitrary exercise of authority and not issued upon reasonable and probable grounds.[1]

Notwithstanding that it was over three years since they were first alerted to the problems of lead contamination, the provincial authorities were hopelessly ill-prepared for the legal challenge. They had only tentative blood sampling results, no phylotoxicological results, and few air sampling results. By contrast, the Institute of Environmental Studies at the University of Toronto had completed a much more extensive testing program, only to find its results ignored by the government and attacked by the industries.

At the court hearing, the lawyer for the refiners called as his witness one of the medical authorities normally retained for such matters by ILZRO. The government had no medical evidence to offer in rebuttal.

Because of the paucity of available evidence, it was not surprising that the presiding judge was unconvinced of the alleged severity of the extent of contamination.

THE INDIVIDUALISM OF COMMON LAW

The issue of proof of the existence or non-existence of an environmental health hazard continues to plague rational discussion of an environmental issue. The question of which party bears the burden of proof, that is, the accuser or the accused, is a legacy handed down to us by our common law heritage. The common law has traditionally favoured after-the-fact compensation of victims and has never developed adequate concern for prevention of harm. It has approached human health as a matter of economic loss and therefore seeks to compensate victims in monetary terms. The plaintiff or victim has the burden of proving that the defendant or wrong-doer is responsible for the injurious act and is therefore responsible for paying the monetary damages. The burden of proving the causal connection between the act and the harm caused by the act is most difficult, if not impossible, in environmental issues.

This legacy of the common law ought not to be applicable to matters of environmental health. For most obvious reasons, it is usually impossible to quantify in monetary terms the extent of the damage caused to the environment, and while practice and experience has dictated the damages awarded for a whiplash injury or a broken bone guidelines are not available to ascertain the cost of the public pain and suffering associated with the pollution of a great river or a large land mass. Even if it were possible to quantify the amount, it would probably be an academic pursuit since few individuals, or indeed industries, would have the financial resources available to pay the amount awarded.

The more important criticism of the common law is that it accepts as inevitable that tortious acts will harm certain people and that it does not set, as a priority, a legal system that attempts to prevent harm from occurring. This tradition is rooted in the social attitude prevalent in England during the industrial revolution. It accepts, without serious question, the assumption that there must be victims in our society and that the best we can do for them is to compensate them after they have suffered harm. Yet even to succeed in a lawsuit for damages based upon the tortious conduct of the defendant, the plaintiff must prove causal connection between the defendant's acts and his harm. In the case of widespread and diffuse emissions, this causal link may be impossible to prove. In any case, it may well be beyond the means of any private individual to retain sufficient technical expertise to assist him in his quest. Finally, can a potential plaintiff conclusively satisfy the burden of proof without producing a victim of the environmental contamination of which he complains? I suspect not. For all of these reasons, our common law tradition has not served us well in the prevention of environmental harm.

The burden of proof must therefore be removed from the shoulders of the potential victim and placed upon the alleged polluter. This is a complete reversal of all that we hold sacrosanct in our legal system; but this suggestion is not only fair, it is also sensible. The alleged polluter has available to him the information and expertise to disprove the allegations. He need not disprove it beyond any doubt at all, but only disprove it on the balance of probabilities. Furthermore, this reversal of onus of proof exists in certain legislative enactments and primarily in the Public Health Act of Ontario.

The dismissal of the closure order was followed rapidly by assurances from James Auld, the minister of the environment, that his ministry would collect the necessary evidence to support a further closure order and that the affected public could be assured of his government's concern. In order to obtain the necessary evidence, an interdisciplinary task force, known as the Working Group on Lead, was formed. This group was the first serious attempt by the Ministry of the Environment to study the nature and location of lead emissions, the extent of degradation in the surrounding air, soil, and vegetation, and the relation of such results to the tests performed on surrounding citizens and particularly on children.

The Working Group on Lead reported its results, conclusions, and recommendations in September 1974, eleven months after its creation. The report pointed accusingly at the sources of so-called fugitive emissions. It conceded that the earlier recommendations of the University of Toronto group had been scientifically valid. Most importantly, it recommended that the standards and criteria for air-borne lead be substantially tightened.

Notwithstanding the lengthy passage of time, the affected citizens were heartened by the contents of the report. At least, the provincial authorities had completed a full scientific enquiry into the extent of lead contamination. The analysis indicated the severity of the problem in the areas of the smelters; the contribution of other possible sources was minimal.

While the Working Group on Lead was preparing its report, the Ministry of Health appointed a three-man task force to examine in broad terms the effects on human health of lead contamination. This task force, called the Rocke Robertson Committee after its chairman, held informal meetings with various medical experts, hoping to arrive at a consensus with respect to the points in contention. The basic argument resolved around the effects of long-term exposure to relatively low levels of lead in the ambient air, the effect, if any, of inhaled rather than ingested lead, and the relative contribution of other possible sources such as paint, pottery glaze, plumbing, and plaster.

The report issued by this committee was a major disappointment to the citizens. While advocating minor reforms, the report equivocated on all the major questions. With respect to the central issues, the report did nothing more than summarize the opposing views. The authors failed to report the fact that experts in the area of lead contamination were split into two schools of thought and that one school of thought was funded in whole or in part by the lead-zinc industry. There was a substantial disagreement between the two schools of thought over the mounting evidence that prolonged exposure to low levels of lead in the air can give rise to subtle neurophysiological changes. The committee failed to specify whether the scientific studies on which they had based their recommendations had been properly presented to the scientific community for peer evaluation either at official conferences or in refereed scientific journals.

Most importantly, however, the committee failed to grasp the fundamental responsibility of the scientist in the area of public health. Faced with admittedly inconclusive evidence relating to the possibility of widespread harm, the committee ought to have opted in favour of a course of action overtly conservative and concerned with the protection of public health. The burden of proof is not upon the unsuspecting members of the public but upon the polluter. He must prove that his activities cause no harm, and unless he can do so his activity ought to be severely restricted or disallowed. There is no room for compromise in such matters unless one views public health as something to be traded off in favour of economic or political gain.

MINISTRIES AT ODDS

In view of the eminent qualifications of the members of the Rocke Robertson Committee,[2] their report was a major disappointment to those who had hoped for a quick ending to the lead controversy which had by then lasted for over four years. But the report answered few questions and created some new issues. While stating that there were insufficient scientific data to permit the setting of sound emission standards, the authors recommended reducing exposure levels. The failure to propose emission standards rendered the entire exercise academic.

The general effect of this report was to undercut the sense of urgency present in the report prepared by the Working Group on Lead. Faced with the differing recommendations contained in the two reports, the Davis government ordered public hearings into the whole question of the extent of lead contamination and charged the hearing tribunal with the task of producing a set of recommendations of its own. In fact, the provincial government decided that, since it could not continue to ignore the problem of lead contamination, it ought to keep the problem under active 'study' long enough to drive it off the front pages of the newspapers.

Perhaps at this point, some discussion is in order about the relative roles of the Ministry of Health and the Ministry of the Environment. When discussing the role of the Ministry of Health, one must always bear in mind the near-total failure of this branch of government in dealing with environmental health problems. The same pattern of failure emerges whether we are talking about lead or mercury, silica, radiation, or any other toxic environmental contaminant. While it is likely that the problems are impossible to prevent, they are not impossible to anticipate or rectify. The scientists in this branch have impressive credentials, but in every case in which it has been involved this ministry has served only to heighten tensions and to diminish public confidence in its work.

The pattern of response is the same: initially, the ministry denies the existence of the problem or states that the problem is under its control. It then refuses to release any incriminating evidence, even to the affected citizens. Finally, it disputes the validity of test results, if those tests were performed by others outside the ministry. The ministry usually adopts the public posture that any information it might release would be dramatized by the media causing public panic based on ignorance. The opposite is generally true. By withholding relevant data, the Ministry of Health gives credence to wild and misinformed rumour.

The Ministry of the Environment is a junior ministry in terms of prestige and power within the cabinet. Its jurisdiction in the area of environmental health is unclear. It must depend upon the Ministry of Health for technical expertise relating to the health effects of environmental contaminants.

There appears to be a severe jurisdictional gap between the two ministries. While the Ministry of the Environment has no medical expertise, the Ministry of Health has no environmental expertise. When the two subject areas overlap, the people of Ontario have no effective agency to deal with the problem.

The public relations of the Ministry of the Environment are somewhat better than those of the Ministry of Health. Generally speaking, information is more readily available and test results are never withheld. Perhaps most importantly, the Ministry of the Environment no longer underestimates the genuine concern of the public over environmental degradation.

THE HEARING BOARD

In November 1974 the government of Ontario announced that it would defer taking any steps to implement the recommendations contained in the reports by the Working Group on Lead or by the Rocke Robertson Committee until a public review of the whole issue had been conducted by the Ontario Environmental Hearing Board. This announcement was received with public dismay, for it meant a further delay before action would be taken. Some five years after the problem was brought to its attention, the government would still manage to have the matter under consideration.

The Environmental Hearing Board conducted a public hearing over a period of some forty-six hearing days producing a transcript of approximately 7600 pages and referring to some 243 exhibits filed during the course of the hearing. Prior to reviewing the recommendations, it is important to reflect upon the hearing process itself, since the Environmental Hearing Board is now the Environmental Assessment Board and the lead hearings acted as a test run of what one might expect in an environmental assessment hearing.

While one cannot dispute the thoroughness of the hearing process, it is interesting that the board failed to involve the affected citizens in the hearing process in any effective way. While all the formal requirements for public participation such as notice and access to information were complied with, the hearings themselves were conducted in a formal and somewhat legalistic manner which served only to heighten the smouldering resentment the citizens harboured against the government. The citizens perceived that their interest, concerns, and frustrations were of less importance to the board than the interests of the larger and wealthier parties – the government and the lead industry. As a result, the citizens, after a few days of hearings, walked out of the hearings and refused to testify. While there have been suggestions that this action was intended to embarrass the Hearing Board and the government, I personally believe that the Hearing Board did not display the necessary sensitivity needed to involve the citizens effectively in its process.

Given the well-funded and professional presence of the lead industry and the government, the board ought to have been particularly sensitive to the precarious position of the citizens. Their resources are non-existent, their training often too little, but the legitimacy of their fears, frustrations, and view points must be recognized. A mechanical process for public participation will always fail unless the people believe the board wants to hear from them. This has been the reason for the success of the Berger commission. In my view, it was the reason for the failure of this board to create an adequate participatory atmosphere.

The Hearing Board report issued on 8 April 1976 and publicly distributed in May is perhaps the most exhaustive analysis of the effects of lead contamination on human health ever undertaken. With certain qualifications, the Hearing Board accepted the recommendations of the Working Group on Lead and, most importantly, accepted a lower and revised lead level in the blood as indicative of excessive absorption. The latter issue was the centre of debate between the lead industry, the government, and the citizens.

The Hearing Board report found a direct relationship between the elevated lead levels in the vicinity of the smelters and the air-borne emissions of lead from the smelters themselves. In other words, they pointed accusingly at the smelters as the culprits. However, somewhat disappointingly, the Hearing Board also concluded that the isolated incidents of elevated blood lead levels were probably related to ingestion of lead by children or occupational exposure by adults. In other words, they refused to say that the elevated lead levels in the environment were the cause of elevated blood lead levels in children and adults. It is difficult to understand this conclusion, except to say that once again the citizens were unable to establish the burden of proof cast upon them. In this particular case, the citizens were able to link the emissions and the elevated environmental contamination, but they were not able to prove that the same particles of lead that escaped from the smelters were causing them physical harm. The requirement that the victim must establish the causal connection between emission and harm is both unfair and unnecessary.

The Environmental Hearing Board made a number of technical recommendations which are both lengthy and complex. While generally
supportive of the findings and recommendations of the Working Group
on Lead, they quibble about the stringency of the working group's
recommendations.

The Hearing Board's report contained a number of administrative and
legislative proposals to clarify the confused jurisdictional responsibility
of the various departments and agencies within both the provincial and
federal governments, but apart from the lengthy technical recommendations the Hearing Board failed to answer the public fears and allegations about lead poisoning. Indeed, as the Science Council of Canada
noted in a report on the Toronto lead case issued in October 1977, the
board failed to recognize the need to 'deal with and resolve the real
problems, backed by both publicly stipulated rules of the game and a
large measure of goodwill.'

LOOKING BACK: SOME RANDOM THOUGHTS

After eight years the issue of lead contamination no longer demands
public attention. New horrors like mercury and radiation command
front-page headlines. One wonders what will be next.

Looking back, it may be that, after all the public furore, little was
actually accomplished. While it is true that some slight improvement has
been noted in the levels of ambient lead in the vicinity of the refiners, the
levels still too often exceed the provincial criteria. The Ministry of the
Environment has recently cracked down on the refiners, imposing strict
control orders aimed at restricting the ever-present fugitive emissions.
But the plant life and the soil in the area of the smelters are still grossly
contaminated and they will pose a hazard for generations of children to
come.

With specific reference to the issue of soil contamination, it should be
noted that both the working group and the task force called for all soil of
more than one thousand parts per million of lead to be hauled away and
replaced with clean soil. On the other hand, the Environmental Hearing
Board and the Environment Ministry both decided that a level of three
thousand parts per million was acceptable. In any event, George Kerr,
the environment minister, told the provincial legislature in 1976 that 'the
first priority (arising from the board's report) is that the companies are
responsible for clearing up and removing the contaminated soil.' In fact,
the vast majority of the clean-up cost of the soil is being paid from funds

generated by the provincial lottery, Wintario, and of a total clean-up cost of approximately $83,000, the three smelters within the city of Toronto are responsible for only $28,000, or less than one-third of the total.

Of the remaining 102 recommendations which the Hearing Board report contains, none have been discussed in any greater detail. At the present time, lead emissions in the area surrounding Toronto Smelters and Canada Metal still exceed provincial guidelines although not to the same severe extent as they did in the early part of the 1970s. The three smelters have spent more than $2 million in the acquisition and installation of abatement equipment.

For the citizens, there lingers a resentment and suspicion of government and of bureaucrats. These citizens feel that they have been badly served by their governmental institutions. The legal system is viewed as the plaything of the rich and powerful. Their legislative representatives were only too ready to abandon their cause because it lacked political advantage.

The bureaucrats have learned that they must change their ways in order to earn the confidence of the public by demonstrating genuine concern and by acting promptly and efficiently. Although the Ministry of Health still has not learned this lesson completely, the signs are most encouraging with respect to the Ministry of the Environment.

Perhaps, most importantly, scientific research will be seen officially to be tainted if it is funded in whole or in part by the affected industry. This question now emerges regularly in environmental issues in the United States and recent correspondence in *Lancet*, the British medical journal, demanded that industry-sponsored researchers identify their funding sources. This question becomes more troubling in view of the cutback in funding for public scientific research. Even the unbiased and genuinely concerned scientist may find that he has to seek industrial support for his research. While the source of funding may not necessarily result in biased research, the researcher cannot be oblivious to this aspect of his work when preparing his results. Apprehended bias, even if factually unfounded, may taint the recognition of such research. While there may seem to be traces of paranoia in pointing to such a division in the scientific community, it only recognizes a struggle which already exists between those researchers who are industry-financed and those who are not.

Finally, in any issue of environmental contamination it seems likely that the government's reaction will be to 'study' the matter, without taking positive steps or appropriate remedial action. Because of this well-established response, I doubt whether citizens will be content to await the final recommendations without demanding positive interim steps. The use of a task force, study group, or fact-finding commission has been used as a successful stall not only in the lead issue but also in the mercury and asbestos issues. While such commissions of enquiry perform a useful role in the education of the public, their impact on governmental policy has been negligible. For example, the Ham Commission investigated occupational safety and health problems in the mining industry in 1974 and reported numerous violations of legislation and ministerial orders; the report clearly indicated the failure of Ontario's environmental and industrial safety surveillance mechanism. Yet, judging from the 1976 reports of the situation in at least one Ontario asbestos mill, industrial practices have escaped effective regulation.

The victims of the lead contamination crisis have been medically treated and discharged to their homes where they face the same threat again. While it is true that there were no cases of gross lead poisoning, it is still too early to tell if any of the children will suffer neurological damage as a result of the high lead levels in their blood. If damage is found, it may be too late to take effective legal action.

But if we are to view the environmental issue as primarily a political, economic, and social issue, should the victims necessarily seek compensation from the polluters? For in fact, it is our society as a whole which has permitted the harm to occur. The responsibility to safeguard the public health effectively is collective; it cannot be individual. If pollution is in fact the environmental cost we must pay if we are unwilling to change our lifestyles, ought we not to compensate the victims out of public monies? For just as we compensate the innocent victims of crime, should we not compensate the innocent victims of industry, the innocent victims of our society, the innocent victims of a political system which tries to ignore their welfare in order to maintain the economic, social, and political *status quo*?

NOTES

1 *The Canada Metal Company Ltd et al* v *MacFarlane* [1974] 1 o.r. 577
2 The members of the committee to Inquire Into and Report Upon the Effect on Human Health of Lead from the Environment were H. Rocke Robertson, Donald A. Chant, and Frank A. DeMarco. Robertson is a physician and former principal of McGill University; Chant is a zoologist and founder of Pollution Probe at the University of Toronto; DeMarco is the dean of the Faculty of Engineering at the University of Windsor, Ontario.

5

The underdevelopment of occupational health and safety in Canada

Robert Sass

The area of occupational health and safety in Canada is underdeveloped in at least two distinct contexts. As a normative field of endeavour, it has failed in the pursuit of its fundamental objective of promoting and maintaining the health and safety of workers; as a scientific discipline, it has failed to keep pace with the changing realities of work and the workplace. We examine here this dual underdevelopment, which seems due to the following two interdependent causes:

1 / In an industrial relations context, labour, management, and government all face internal and external 'contradictions' in their attempts to deal with occupational health and safety problems. These contradictions flow from the inconsistent and conflicting priorities that exist within our complex political economy and they tend to hamper the development and implementation of effective occupational health and safety practices.

2 / Much of the 'scientific' knowledge associated with the field is merely an aggregation of myths and folklore disguised as science. The existence of such a body of myths – that is, 'conventional wisdom' – serves further to hamper the development of occupational health and safety both in theory and in practice.

'Contradictions' and 'conventional wisdom' are the key words and, as such, form the hub of this paper.

THE CONTRADICTIONS

Dangerous work is the biggest threat to the health and safety of workers in the workplace! No matter how self-evident this statement appears to be, it is a critical one as it brings to light a key concern – the nature of work and the forces that shape it. In order to introduce the principal contradictions that the field of occupational health and safety faces, it is useful to take an 'industrial relations' perspective. This perspective is doubly desirable as the key actors in the industrial relations arena – labour, management, and government – are the major forces in shaping the nature of work (the role of 'science' will be discussed later). In order to introduce those contradictions that impede the development of healthful and safe work, let us begin by examining the role and perspective of each individual actor in the field of industrial relations.

Management
Given the nature of our political economy, it is clear that most industrial enterprises survive only as long as they continue to provide an acceptable financial return. Management is thus primarily responsible for providing leadership that will ensure the survival of industrial and commercial ventures. This, in itself, makes it necessary for management to develop a strong economic and financial consciousness. This orientation also serves to shape the fundamental contradiction that management faces in its attempts to deal with occupational health and safety problems.

Industrial *safety* is profitable only when the direct and indirect costs associated with accidents (such as production shut-downs, damaged equipment and materials, increased workers' compensation assessments) exceed the cost of eliminating these accidents. From a solely financial perspective, it is often better to allow all but the most blatantly dangerous conditions to exist rather than to incur the additional financial outlay necessary to make the workplace safe. In the mining and construction industries, for instance, the cost of making work 'relatively safe' – that is, as safe as office work – is extremely high.

In the same light, management has even less of an incentive to moderate the negative *health* effects flowing from the workplace, since very few of the costs associated with industrial illness are absorbed by the industry that produces them. Today most of these costs are borne by the victims of industrially induced illness (loss of income and the relevant social and psychological costs) and by taxpayers in general (the cost of medical care, rehabilitation, social assistance, and so on). As industry is allowed to engage in activities that produce illness on a relatively penalty-free or cost-free basis, it has virtually no economic incentive to prevent the health problems that it creates.

Beyond a certain subsistence level, healthful and safe working conditions rarely contribute to industrial productivity and profits. Making the workplace healthier and safer usually involves an increased investment in equipment and processes, a better preventative maintenance program, more shut-downs, a slowing down of the pace of work, and so on. In short, such improvements often impose a financial burden on an industrial firm, and sometimes even threaten its competitive nature or its very existence. In general, management faces a contradiction between health and safety concerns and production and profit priorities. These variables are interdependent and inversely related inasmuch as an increase in the level of occupational health and safety often results in a decrease in production and profitability, and vice versa. Because of this relationship, it is both a mathematical and practical impossibility for management to maximize both concerns simultaneously. This means that trade-offs between the two must be made.

Since the survival of management (or better, managers) is based upon the survival of the enterprise that employs them, and the key to their survival is profitability and production, it is understandable that management, if forced to choose between concerns of production and profit and concerns of occupational health and safety, will often opt for the former. This reflects the contradictory prerequisites for survival found in our political economy rather than a general lack of morality on the part of management. Management then may be in favour of healthy and safe production, but first and foremost they want production.

Labour
Labour's concern for occupational health and safety is currently outstripping its traditional emphasis on conditions of employment such as wages, hours of work, seniority, etc. This concern reflects a growing awareness of both the nature and the extent of the health and safety problems that workers face today.

Many people argue that management and labour are natural adversaries and that competition between them is necessary in order to maintain a 'balanced' state of affairs. Flowing from this is the notion that occupational health and safety are not management concerns unless they are bargained for by labour at the bargaining table. (Exceptions are situations where such conditions are improved unilaterally by management to increase productivity and, of course, management activities in the area based upon legislated standards.) This argument, however, assumes that labour has the ability to fight for occupational health and safety freely and effectively. This assumption is a false one, as the problems continue to outstrip labour's ability to deal with them.

Trade unions face a number of difficulties in attempting to deal with the health and safety of workers. Many unions have become institutionalized and are slow to respond to changing needs. For instance, many individuals within a union's hierarchy are there because of their skills in the traditional bargaining area of conditions of employment; but because they have been away from the shop floor for so long, they are often incapable of understanding and dealing with health and safety problems. Further, the centralized master bargaining agreements used by many unions often compound the problem as they make the enumeration, evaluation, and inclusion of differing health and safety needs of industrial locals a cumbersome task. And even when a union wants to broaden its scope and include health and safety, it is often hamstrung by a lack of appropriate technical and financial resources.

Currently, labour appears to rely heavily upon government for occupational health and safety reform. For example, if one examines the health and safety clauses found in many of the collective bargaining agreements signed recently, it is not unusual to find simply a statement that 'such and such a piece of occupational health and safety legislation shall apply.' (For instance, in the case of Saskatchewan it would be The Occupational Health and Safety Act, 1977.) Often, this is the entire content of the health and safety clauses. To a certain extent, this sort of reliance upon legislation is tactically necessary. If the legislation offers any sort of tangible improvements, the unions are placed in a better strategic position by accepting it and by refraining from bargaining for additional concessions. Quite simply, when they come to the bargaining table they will not have to trade off other factors such as wages, sick leave, etc., in order to attain desired health and safety objectives. The fact that improvements in health and safety conditions often involve a sacrifice in the more tangible and immediate aspects of work (that is, wages) often serves as a disincentive to union activities in this area. If the union membership does not see occupational health and safety as a priority, the executive risks incurring the wrath of the membership if it overemphasizes these concerns.

The labour movement is thus not free to attempt to maximize the health and safety of workers. Besides the problems mentioned previously, labour also faces a fundamental contradiction. As with management, the survival of workers is related to the survival of industry and commerce. Occupational health and safety is not compatible with production and profit and workers often face the threat of plant close-outs or displacement by new and safer automated equipment if they press too hard.[1] Workers may want healthy and safe work, but first and foremost they want work. Consequently, they are likely to accept unsafe and unhealthy working conditions rather than unemployment.

It is a myth that workers who do not like their working conditions are free to go elsewhere.[2] A worker cannot easily transform personal circumstances and skills or break social or geographical ties just to find healthier and safer work. The conflicting relationship between occupational health and safety and job security means a dilemma for workers in attempting to improve their own well-being.

Government
Lately, government has been playing a growing role in the area of occupational health and safety. New legislation has recently been introduced or is in the process of being developed in most Canadian provinces, as well as in the United States and many European countries.

The growing role of government in this area has been attributed to a number of different factors. For example, it has been suggested that increased government involvement reflects a recognition of the imbalance in bargaining power that exists between labour and management which makes the collective bargaining process an inadequate means of dealing with occupational health and safety problems.[3] On the other hand, some feel that legislation is being introduced with the intention of eroding labour's power to pursue health and safety objectives, while allowing industry to continue to function virtually unhindered. Both notions may be correct, depending upon the situation. It is quite clear that occupational health and safety legislation differs from province to province in terms of both its *intent* and its *enforcement* by government.

To suggest, as do some conflict theorists, that governments respond absolutely and subserviently to the desires of the more powerful economic group is an oversimplification. Governmental activities are largely a reflection of generally accepted political and economic priorities as well as of the relative pressures that competing interest groups can bring to bear. However, powerful interests can and do exert pressures on government which prevent it from dealing effectively with occupational health and safety problems. Consider the strength that industry possesses if it is faced with legislation that it considers undesirable. Because its capital is often relatively liquid and easy to transform, industry can realistically threaten to leave the province or country. This type of threat is loaded politically as the loss of industry means a loss in the production of real goods and services and taxes as well as increased unemployment. The term 'runaway capital' indicates that while Canada (or any other country) may need industry, industry does not necessarily need Canada.[4] Government, when faced with the possibility of the flight of capital, has been known either to soften standards or to keep the standards and soften their enforcement. A 'political cost–political benefit' analysis on the part of government in this type of situation often reveals the *political cost* of implementing better health and safety standards to be prohibitive.

Government faces the contradictions that both labour and management face, and more. It must consider the relative value of introducing higher occupational health and safety standards in terms of their effect on industrial development, GNP, and employment. Higher standards will tend to lower the GNP (still considered by many to be a measure of social well-being), act as a disincentive to investors, and produce greater unemployment. Production priorities have acted to the detriment of workers' health and safety and yet it is generally recognized that workers also consider unemployment to be a serious injury. Consumption priorities have also introduced more contradictions, as the general public desires the goods and services that an industry is capable of providing, but may not want the negative health and safety effects associated with the industry such as injured or ill workers and environmental pollution. (The argument that industry does not operate for the sake of its workers' health and safety but for the sake of its many consumers misses the point, however, that most consumers are just workers away from their place of work.)

The fact that government must consider all of these contradictory aspects when making a decision indicates that the simultaneous maximization of both social and economic priorities is currently impossible. This is clearly evidenced in most of the occupational health and safety legislation that exists today; rather than representing an ideal state, it represents the trade-offs government has made in light of the contradictions that it faces.

THE CONVENTIONAL WISDOM

The underdevelopment of occupational health and safety is closely linked to a number of unscientific principles and perceptions which have been allowed to shape the field. Progress and innovation have been blocked by this aggregation of folklore, bent facts, and fiction. The 'conventional wisdom' has, in turn, evolved through and been given credibility by that group of individuals commonly known as the 'experts,' the professionals, the specialists, the scientists. The actions of the experts themselves have been a response to two powerful forces that they themselves face – the manner in which 'professions' are organized, and the contradictions that the experts face in relation to our political economy.

Most professions are organized in a manner that encourages specialization, and although specialization does have many positive features it has also led to the development of a number of narrow and competing 'single factor–single cause' frames of reference amongst the experts. For example, when they are confronted with an occupational health and safety problem, they tend to define the issue in terms of their own narrow area of expertise and, by doing so, define the solution as being one that only they are capable of administering. Common sense and interdisciplinary discussion are often sacrificed during this process; instead, each group of experts will tend to overstate its own individual usefulness while failing to come to grips with the complexities of the problem. Consequently, the experts seldom see the issue in its entirety, since any actual issue generally fails to divide and/or simplify itself in order to accommodate their narrow and often competing perspectives.

The manner in which the experts are trained is also problematical. The intense training they receive in their area of expertise tends to make them averse to new and different ideas. And the fact that most 'scientific truths' are subject to revision and change puts many of the experts in a state of continual flux between elitism and insecurity. For instance, while they may see their field as a 'priestly craft' in untroubled times, if their activities or theories are questioned by laymen they will quickly band together for mutual support. This huddling is usually accompanied by the mumbling of mystical incantations developed by the profession for just such an occasion – the purpose of these abstract and verbose meanderings being to make the layman feel like a village idiot. (It is interesting that most people, when unable to understand an expert, will attribute their bewilderment to the expert's 'genius' and, when unable to understand a layman, will attribute it to his or her stupidity.) Not surprisingly, this tactic is often successful as the layman, in confusion, often feels incapable of continuing to question the experts.

The experts' actions are also hindered by factors external to the way the professions and disciplines organize themselves, as the priorities found in our political economy also influence their perceptions and activities. For example, their employment in industry is dependent upon their ability to contribute to organizational objectives such as production and profit, and workers' health and safety may suffer because of this. Even when the experts are not employed by industry, they tend to deal with occupational health and safety problems from an outmoded economic perspective in which economic well-being (that is, productivity, profitability, GNP, etc.) is considered to be a meaningful indicator of social well-being. Consequently, the priorities of our productive system are normally taken as given and most of the attention is focused on the nature of the worker and the problems associated with getting the worker to fit into the workplace. This, in my estimation, is a backward approach as it subordinates the worker to the productive system. It is more appealing to envision a productive system that is designed to fit the needs of workers as well as those of consumers. In general, the experts will not question the nature of work itself; therefore, they become incapable of understanding the interaction between the worker and the workplace. The occupational health and safety implications flowing from this blindness are enormous.

By not questioning the nature of the work, the experts have little option but to develop a curative approach towards occupational health and safety instead of a preventative one. In essence, they tend to try and patch up the damage done to workers after it occurs.[5] Workers are too often the guinea pigs that prove the dangers of industrial processes and substances. While this sort of approach can only continue in the absence of widespread outrage, outrage is blocked by the experts who will often deny that a given problem exists, no matter how real it is.

The experts have also played a significant role in the development of myths that place the blame for occupational injuries and illnesses upon those workers unfortunate enough to incur them. 'Blaming the victim' is a tactic that currently manifests itself in such unscientific concepts as 'accident proneness' and 'employee carelessness.'[6] Blaming the victim and the emphasis on curative measures together form a substantial part of the conventional wisdom associated with occupational health and safety today.

Up to this point we have discussed the pressures and barriers the experts face, their shortcomings, and their role in developing the conventional wisdom in broad general terms. In order to fortify these observations it is appropriate to examine those professions and disciplines associated with the area of occupational health and safety at an individual level.

The medical profession
Medical skills have often been deployed in a defensive capacity to control the incidence of blatant illness or to patch up injuries rather than

to improve the level of worker health and safety. Historically, before
and during the early years of the industrial revolution, the medical
profession's primary task in the area of occupational health and safety
was to treat and fence in diseases and epidemics associated with the
workplace. The inhuman working conditions made workers more susceptible to cholera, tuberculosis, fevers, etc., and the fear that other
classes might be affected through contagion led to attempts to confine
these diseases to the workplace.[7]

It was in this manner that our public health model was originally
established. It was merely an extension of middle-class private medicine
and, consequently, much less attention was paid to healthy working
conditions than was paid to the needs of the upper and middle classes.
Today, we can see an interesting parallel to this situation if we consider
the problem of environmental pollution.

Many people are concerned about the hazardous effects on the general public of the pollution which is spilling out of the workplace. When
scientists are queried as to what extent environmental pollution could be
reduced by eliminating the health hazards found in the workplace the
answer is invariably 'considerably.' This has led many to suggest that
occupational health is merely a sub-category of environmental or public
health. This argument ignores what history has taught us about the class
origins of public health.

At present, the public outcry is more intense regarding the
'neighbourhood effect' of pollutants such as asbestos, air-borne lead,
sulphur dioxide, radiation, etc. than with occupational hazards *per se*.
Many environmentalists want a safe environment, but still desire the
fruits of modern technology. This sort of perspective and current public
health priorities could again lead to the enclosure of hazardous situations and substances within the confines of the workplace and, therefore, provide little or no protection for workers. If we reject the public
health approach, however, and direct our attentions specifically towards
the worker and the workplace, we can positively improve the level of
worker health and safety and also provide positive health benefits to
society at large.

The application of the public health model has contributed greatly to
the underdevelopment of industrial medicine as a science. Neither
medical schools nor the practice of medicine adequately equips the
physician with the knowledge and techniques needed to deal with occupational health and safety problems. In Canada there are no certification
procedures for doctors practising occupational medicine and no insti-

tutes of higher learning have any profound capability in this area. All that currently exist are a few short courses on certain aspects of occupational health and safety. Consequently, the 'industrial physician' is usually merely an individual who practises medicine in the factory instead of on Main Street.[8] The shortcomings of this type of approach are further compounded by the problems flowing from the manner in which the medical profession in general operates.

The structure of the medical profession usually means that the physician sees the individual divorced from his or her work environment; therefore, most physicians have problems recognizing the relationship between occupation and illness. For example, as workers in a given workplace or industry usually have many different doctors, the medical profession has difficulty identifying specific health and safety problems which are directly attributable to a particular work process or work environment.[9] Even the company doctor seldom goes out onto the factory floor; instead, his patients are brought into his sterile office. These factors lead the physician to see silicosis and asbestosis as problems of the lung and not as problems of hard-rock and asbestos mining. Hearing loss is seen as a problem of the ear and not the pipe mill. This narrow concentration on the physical aspects of the worker has led to the development of many myths, both in the past and at present.

For years, company doctors denied the existence of black lung disease, and in some instances coal miners were even told that a little coal dust was good for them. Members of the paint trade were informed that lead poisoning was caused by unwashed hands (somehow the effects of air-borne lead were totally ignored) and the high incidence of dermatitis amongst workers working with coal tar and similar substances was explained in terms of poor worker hygiene or washing oneself *too* much (this, of course, depended upon the doctor). Employee drinking was used to explain away the high incidence of employee illness in the chemical industry. These are all cases of 'blaming the victim.'

Today we see continuation of this non-scientific type of thinking in cases where employee illness is blamed on improper handling of substances, employee carelessness, and so on. The medical profession's narrow focus does not allow those involved to diagnose properly or prevent the problems of the workplace – they usually confine their activities to treating these problems only after they manifest themselves in the worker – and, yet, their ignorance regarding the nature of work and the workplace does not stop them from making judgments and decisions they are realistically incapable of making. (One medical officer at a provincial workers' compensation board stated that the area of industrial diseases was 'out of his element.')

The judgment of many physicians, and their inability to act in the interest of workers' health and safety, is further influenced by a number of conflicts of interest that they face. The whole question of disclosure of occupational hazards provides a case in point. Consider the position that most company doctors find themselves in if they uncover a health or safety hazard in the workplace. While it is true that they can bring these hazards to the attention of management, they are not in a position to push for changes in the work environment or to inform workers of the hazards they face. These are considered to be managerial prerogatives and, due to incompatibility between workers' health and safety and company production and profits, management often responds with silence and inaction.

Physicians can also hide their failure to disclose hazardous situations to workers and other concerned parties behind the cloak of confidentiality. In certain instances, workers are denied access to their own medical records through the use of the confidentiality ploy. There have been numerous cases where a physician (not necessarily the company physician) has given in-plant or on-site examinations to workers, turned the results over to management, and denied workers the right to see their own results.[10] Physicians often rationalize such actions by professing a concern that patients may misunderstand the results of their tests and worry or, in the case of terminal conditions, that the individual may be better off not knowing that he or she is going to die. (The fear that a patient may misunderstand the situation is usually an offshoot of the mystification found in most professions and the notion that workers are 'village idiots.') However, as one doctor rather candidly stated, 'confidentiality often has more to do with one's own protection than with the best interest of the patient.' Confidentiality often means that the doctor is the only one who really knows that a problem exists and that this information is kept from the victim or anyone else in a position to take corrective action.

Even in the medical research field, occupational hazards are often obscured. Much of the findings come from industry and this in itself tends to influence the selection of subjects, the methodology employed, and the final outcome of the studies. For example, the Industrial Health Foundation was commissioned by the American Textile Association in 1970 to do a study on a lung disease, caused by cotton dust, called byssinosis. The IHF study was a model of equivocation; IHF spokesman Dr Robert de Traville stated that byssinosis is 'best described as a "symptom complex" rather than a "disease" in the usual sense. We feel that this term may be preferable, first, in order not to unduly alarm workers as we attempt to protect their health, and secondly, to help avoid unfair designation of cotton as an unduly hazardous material for use in the textile manufacturing industry, raising the fear that engineering control of it may be costly and that it may be better, therefore, to switch to some less costly material ... It would be unfortunate indeed, if we were to discontinue use of cotton fibres unnecessarily as a result of not being technically sound in our discussion of potential hazards and controls ... It would be wise to delay temporarily the extensive and expensive changes in plant equipment and procedures aimed at dust suppression alone.'[11]

What De Traville was advocating, in essence, was a delay in dealing with a problem that has already cost between 12 and 30 per cent of the 230,000 US textile workers the normal use of their lungs.[12] This approach also illustrates a tendency on the part of many physicians to avoid saying whether or not they believe a given substance is dangerous until they completely understand its etiological and pathological effects.[13] It ignores the common-sense method of observing the relationship between the use of a substance and the existence of certain common health problems and has led to inaction costing many workers their health, and in some cases their lives.

The legal profession
Our legal system has never been particularly sensitive to the health and safety needs of workers. During the late 1800s and early 1900s there was a noticeable absence of those legal foundations necessary to prevent occupational illnesses and accidents. This shortcoming was compounded by the legal barriers that workers faced if they tried to get compensation for an injury they received in the workplace.[14] Severe criticism began to emerge as workers were excluded from the ordinary protection of the law that was available to the general public. This criticism, along with the fear expressed by industry and commerce that the courts might become more generous in their assessment of damages in the case of employer negligence, led to the introduction and acceptance of workers' compensation schemes.

Workers' compensation became a no-fault and after-the-fact means of awarding compensation to victims of industrial accidents (and, more recently, a few specified industrial illnesses). Although these schemes put workers in a relatively better position than before, in the event that they were injured, it was industry that benefited the most. Firstly, industry found that paying their compensation assessment was usually cheaper than introducing preventive measures into the workplace, and, secondly, workers covered by a compensation scheme had to forfeit their right to sue their employer for negligence. For industry, workers' compensation became an effective means of minimizing the costs associated with injuries and illnesses.

The shortcomings associated with workers' compensation schemes were widely recognized, but it was felt that these defects could be met through the introduction of legislated standards and codes related to occupational health and safety. Unfortunately, the introduction of such regulations and codes has never been any guarantee that they would be adequately enforced.

The enforcement of legislated health and safety standards falls under the area of criminal law, and judges are often reluctant to label as criminal something which is perceived as being merely inefficient (that is, as opposed to being malicious). Therefore, if an employer is prosecuted for failure to comply with safety regulations in a case where no damage has been done, there may be a tendency to view him as the victim of a bureaucratic process. A recent case of this type in Saskatchewan illustrates the point. When the prosecuting lawyer was asked, 'What is the problem? Why the concern if no one was hurt?' he explained that the concern was to try to prevent accidents *before* they happen. The employer was subsequently fined less on each count than someone who is caught speeding on the highway.

In general, lawyers receive an education that tends to emphasize case law (very little of which is applicable to occupational health and safety) as opposed to statutory provisions, regulations, and regulatory techniques. The training of lawyers and their experiences in legal practice tend to shape a perspective that emphasizes the consequences of an injury rather than preventative measures. Also, lawyers and judges are trained specifically in law and often have little opportunity to develop an appreciation of the nature of work.[15] Because of this, it is difficult to present a clear impression in court of the hazards involved in a given work situation. In the ordinary courts there is also a general reluctance to view the actual scene of the hazard, injury, or fatality, as this is not considered to be a normal part of criminal proceedings.

Because of these and other problems, some individuals have questioned whether the courts are an adequate forum for dealing with occupational health and safety problems – particularly from the aspect of prevention. This, in turn, has led some to suggest that there is a need for an occupational health and safety tribunal similar in nature to the Labour Relations Board, composed of persons who have an understanding of industrial life.

Industrial engineering
Industrial engineering has largely been shaped by 'Taylorism' or 'scientific management.' Taylorism was introduced in the early 1900s when Frederick Winslow Taylor, an engineer, and Frank B. Gilbreth, an early proponent of time and motion studies, developed principles designed to expedite industrial production by combining the 'one best way' of doing things with the technology available at the time. The key to the success of Taylorism, as Taylor saw it, was the necessity that management dictate to the worker 'the precise manner in which work is to be performed.' Consequently, rather than being scientific as the name implied, scientific management was merely a collection of techniques designed to control the worker and the production process.

Taylorism, as applied in the field of industrial engineering, has resulted in the de-skilling of workers since it has advocated that worker skill, an intellectual quality, be systematically removed from the shop floor. Conception of the productive task and control over it has largely been moved into the hands of management or built into the production process itself leaving the workers as mere executors of pre-determined tasks. The possibilities for craftsmanship have been lost as more and more workers have become involved in machine-tending and assembly-line type work. In this manner, management has not only been able to control production and workers, but it has also been able to justify paying lower wages for 'easier' (that is, unskilled) work.

The introduction of the principles of Taylorism has led to increased specialization, standardization, and the minute subdivision of labour in the workplace. Work has been transformed into jobs and workers have been forced to contend with the accompanying fatigue, alienation, monotony, and inattentiveness. This situation has not only introduced new and undesirable psychological factors into the workplace but has also had an effect on safety. (It should be remembered that the concern for occupational health and safety is a concern not only for the physical but also the psychological and social well-being of workers.) Monotony and boring repetition leads to inattentiveness which in turn can lead to a hazardous delay in the reaction of workers to dangerous situations. Speeding up production through either quickening the pace of the work or introducing payment incentives such as 'piece rate' can lead to a conflict between safety and production on the part of the worker with production often finally taking precedence.

It is interesting to note that many individuals continue to attribute accidents that result from these factors to 'employee carelessness.' The implication of this notion is that workers have a large measure of control over their work environment and are thus able to act independently and in a manner consistent with their own safety. The continual removal of control over production from the shop floor and its confinement in the hands of management and the production process make this contention unrealistic. Rather than providing an answer, the 'carelessness' argument merely raises the question, 'Why do workers "care less" about their own health and safety?' The answer, it seems, revolves around the whole question of control. The way work is organized often does not allow workers to act in a safe manner.

Unfortunately, many industrial engineers appear to ignore the influence that the technical aspects of work have on workers' behaviour. W.H. Heinrich, a noted industrial safety engineer, suggested that of all the accidents he investigated in a particular study 88 per cent were due to 'man failure' (the unsafe act), and 10 per cent were due to 'machine failure' (the unsafe condition), and 2 per cent were unpreventable (acts of God).[16] An examination of Heinrich's study indicates its overly simplistic nature. It is clear that one cannot assign a single discrete cause to most industrial accidents and yet the results of his study have been widely accepted. This acceptance has had negative results as it adds fuel to the 'employee carelessness' argument and draws attention away from the hazardous nature of the work environment. Safety engineers and management using this approach have reasoned that it is often more economical to attempt to develop 'safety consciousness' in workers instead of making the substantial financial outlay necessary to correct any but the most hazardous working conditions. The results of this approach from the standpoint of workers' health and safety has been poor, since 'safety consciousness' is a necessary but not sufficient condition for safe work.

In general, the training that industrial engineers receive is aimed at providing skills that will assist industry in meeting its production requirements. Because of this and because of their dependence upon industry for employment, their involvement in the introduction of occupational health and safety innovations is usually limited to those instances where such innovations will coincide with industrial production and profit priorities, or where they are legislated. Industrial engineers usually perceive the worker and the workplace as being distinctly separate and are therefore unable to understand the health and safety implications that flow from the interaction of the two.[17]

The hygienists
The development of industrial hygiene has been a direct result of growing industrialization. During the First World War the labour market was tight and in key industries such as mining, chemical processing, and metal refining industrial production was largely dependent upon worker productivity. As industrial hazards were abundant and injured or ill workers could not be readily replaced it became necessary to develop techniques to protect the health and safety of the workers. It is clear that this was not done strictly for humanitarian reasons. At that time industrial engineers and industrial physicians were often operating at odds and the need for some sort of professional interaction between the physical and technical aspects of work led to the development of industrial hygiene.

Early hygienists often operated by the 'seat of their pants' as no professional standards or guidelines existed at the time (they could come from almost any discipline as long as they had some knowledge in the areas of health and industrial technology). The 'seat of the pants' approach often consisted of such things as giving workers a damp cloth to put over their mouths to keep them from inhaling large particles (inhalation of the particles often caused time-consuming coughing and retching) or rotating workers through work areas where the exposure to hazardous substances, such as lead, was quite high. This hit and miss approach often resulted in merely altering the original problem instead of solving it. In many cases, production commitments were of utmost concern and the approach taken often digressed to one of simply treating symptoms as opposed to getting at the source of the problem.

The profession of industrial hygiene was 'legitimized' in 1939 when the American Industrial Hygiene Association was officially formed. Its constitution stated that 'Industrial hygiene is that science and art devoted to the recognition, evaluation, and control of those environmental factors or stresses arising in or from the workplace, which may cause sickness, impaired health and well-being or significant discomfort and inefficiency among workers, or among citizens of the community.'[18] Up to the present, however, industrial hygiene has concentrated on the 'recognition' and 'evaluation' of industrial hazards, due to its relative inability to exercise much 'control' over such hazards.

The role of the hygienists has largely been defined by the perceptions of the medical profession and by the needs of industry (which employs most hygienists). Hygienists generally monitor and control substances in ways that minimize the cost of compliance with existing legislation, and solve occupational health problems, but do not prevent them. Much of their training is geared towards solving technical problems (for example, putting heat shields around motors) and not towards preserving the health and safety of workers.[19] They have become an integral part of the administrative problem-solving team and their role as such is inseparable from their dedication to the success of business. As has been concisely stated, 'Any industrial hygienist who does not apply his talents to the best interest of his employer is short changing that employer. His importance to his employer and his status in the industry with which he is affiliated is dictated to a great degree by his measureable value to that industry.'[20] Because of the nature of their training and of their relationship to industry, it is often difficult for hygienists to act in the best interest of workers. Their role in establishing 'threshold limit values' (TLV s) and 'maximum allowable concentrations' (MAC s) illustrates this well.

The development of TLV s has been seen as a major role of industrial hygiene – particularly with the advent of more precise scientific instrumentation. The notion behind TLV s is that almost any substance entering the body or coming in contact with it will be injurious at some level of exposure and tolerated without effect at some lower level. While TLV s are supposed to be set to reflect a 'tolerable' level of exposure, certain problems exist.

It is often extremely difficult to estimate accurately the cumulative effects of long-term exposure to a given substance before many of the negative effects begin to manifest themselves in workers. (Worker health is invariably the final proving ground for any TLV.) Also, almost nothing is known about the synergistic effects that result when workers are exposed to two or more substances in combination, or to one or more substances operating in conjunction with a given work environment. Hygienists have tended to concentrate on the effects of a given chemical or substance in isolation both from other substances and from the workplace. Unfortunately, when TLV s are established they are often seen as 'scientific absolutes' and, as such, induce a false sense of security. Even those TLV s introduced as crude approximations or stop-gap measures take on an air of 'scientific' respectability until it becomes clear that they are too high and need to be adjusted downwards.[21]

The technical problems that hygienists face are not the only ones that deny workers protection from the hazardous substances found in the workplace. There are also certain practical problems restricting their ability to act in the best interest of workers' health and safety. First, hygienists face a conflict of interest when it comes to proposing acceptable levels of exposure to industrial substances, for it is undesirable for them 'to bite the hand that feeds them.' Second and perhaps most important is the fact that hygienists do not legislate TLV s or MAC s – this is done by government. Instead, hygienists are often simply used by industry to give scientific respectability to arguments directed at government in favour of increasing or reducing TLV s or MAC s.[22] Hygienists are often the first to admit that 'acceptability may be more important than toxicity in deciding what concentrations may be present in the working place.'[23] In reality, TLV s and MAC s are usually set by government only after the impact associated with their introduction is considered to be both economically acceptable (that is, in relation to industry) and politically feasible. This, in turn, may explain why TLV s exist for only about 500 of the estimated 15,000 toxic substances used by industry today.

In essence, industrial hygienists have neither the ability nor the opportunity to exercise any independent influence. Their role as objective professional scientists has slipped to one of being technicians; and instead of becoming leaders they have become followers who take the nature of the workplace as given and attempt to make incremental and curative improvements within this framework.

Psychologists and psychiatrists

Psychologists and psychiatrists have directed their attentions primarily towards occupational safety, not occupational health. Their efforts, for the most part, have revolved around attempts to determine the existence and nature of certain psychological characteristics within the individual making him 'prone' to accidents, industrial or otherwise.

A survey of the literature indicates that most psychologists find it necessary to attribute accidents to personal factors and attitudes alone. For example, many in the psychoanalytic school of thought contend that a great many accidents result from a subconscious desire on the part of individuals to punish themselves for, or to rebel against, feelings of guilt, fear, and anxiety stemming from early childhood experiences.[24] Not surprisingly, they also contend that the only effective means of accident prevention lies within their own area of expertise – psychotherapy. Many psychologists, on the other hand, in keeping with their narrow focus on personal factors, tend to explain accident causation in terms of bad attitudes, resentment towards authority, inattention, fatigue, maladjustment, etc., on the part of workers. (Again we can see a classic example of 'blaming the victim.') Unfortunately, this sort of analysis ignores the role that the workplace plays in causing accidents.

A study done by W. Kerr illustrates this point well. Kerr suggests that some individuals will have more accidents than others because they are 'unable to adjust' to the psychological and physical stresses found in their work situations. Complementing this, he added that workers who are unable to participate in setting work goals will show a lack of judgment in their work creating inattentiveness and a propensity towards accidents.[25] While Kerr's analysis explains accident causation in terms of the worker, it can be turned around to make a completely different point.

If a worker is responding to the work environment with behaviour that is leading to accidents, then it is realistic to suggest that he is *reacting* rather than acting independently. Consequently, it is possible to explain accident causation not only in terms of unacceptable psychological responses but also in terms of unacceptable cues to action found in the workplace. The fundamental question then revolves around determining whether the behaviour that contributes to accidents is a deviant action or a normal human reaction to an abnormal work situation.

As with many behaviouralists, Kerr neglects to question the nature of work and its role in shaping human behaviour. Industrial psychologists and psychiatrists have very little incentive to understand the nature of the relationship between a worker and his or her work (as with many professionals, there appears to be a pronounced failure on their part to understand industrial life). Often they are hired by industry merely to help increase productivity by helping to improve worker morale or to detect workers with non-adjustive (that is, less productive) behaviour for the purpose of either helping them or weeding them out. Implicit in the approach is the notion that it is desirable to shape the person to fit the machine or process and to detect those who cannot or will not adjust themselves.

Rather than assisting in accident prevention, many psychologists and psychiatrists have instead provided pseudo-scientific arguments that allow industry to shift the blame for accidents from the workplace to the worker. The whole notion of employee accident proneness is a good example. Although investigation into this area began in the early 1900s, accident proneness has yet to be scientifically defined or proven to exist.[26] Consequently, a psychological profile of the accident-prone individual has failed to emerge for the purpose of detection and prevention. Instead, all we have is a circular argument which suggests that a person who has an undue number of accidents is accident-prone (or careless) and goes on to define accident-proneness (or carelessness) to mean 'having an undue number of accidents.' Even though a number of articles have been written which show the simplistic and unscientific nature of this concept,[27] it is still widely used to explain the occurrence of accidents. This is indicative of the general tendency of psychologists and psychiatrists either to overlook or to fail to acknowledge the role that social, technical, environmental, and organizational factors play in accident causation.

Economists
Economic theory, for the most part, has not been compatible with the health and safety needs of workers or society in general. The focus of economic theory has been based on the survival of the firm through profit maximization (micro-economics) and on aggregate national and international economic activities (macro-economics). Together these bodies of theory have served to justify the pursuit of profit, industrial growth, and growth in GNP, while either ignoring or downplaying the negative social effects that accompany such activities. Today, it is becoming quite clear that increased industrial activity has widespread health consequences, and that growth in productivity and GNP are not indications of increases in the general level of social well-being.[28]

Economics, while considered by many to be a social science, generally either ignores or assumes away the complex human aspects of economic behaviour. For example, labour is considered a factor of production and the individual is considered simply as bent on maximizing his or her own self-interest. This restrictive and simplistic perspective both dehumanizes the worker and artificially separates economic activities from the realm of political and social action. This abstraction from reality has often provided a misleading picture of the nature and effects of economic activities. A look at classical economic theory illustrates this point well.

Adam Smith suggested that the supply price (wages) for labour would take into consideration dirty, dangerous, and unhealthy work. These conditions would be accompanied by increasing wage costs and the introduction of 'hazard pay,' both of which would act as incentives for industry to correct unsafe and unhealthy working conditions. This theory, of course, has seldom been backed up by fact as much of this dirty and dangerous work has also been the lowest paid. What has happened in North America is that, when the need arose, immigration policies were loosened up to encourage an influx of cheap immigrant labour to perform this type of work. Historically, the steel industry and the building of railways in both Canada and the United States provide good examples of cases where this has occurred.

In all fairness, it must be noted that economic theory has not stood completely still since its early development, though much of the out-dated theory is still used by industry and commerce to justify their actions. Today, welfare economics informs us that many industries impose damage or costs on society that the industries themselves do not have to pay for (in economic jargon these are called social costs, or external diseconomies, or negative externalities). Instead, the victims of the damage and society in general end up absorbing these costs and are, in effect, subsidizing these industries. What is suggested, therefore, is that the cost of this damage be calculated and passed back to those responsible. This, in the case of damage to the health and safety of workers for example, would theoretically provide an incentive for industry to reduce the damage flowing from the workplace, or at least raise the price of the goods produced by the guilty industry to reflect their true cost.[29]

The social-cost approach, while being intuitively appealing, is difficult to apply for a number of reasons. First of all, at present it is technically impossible to determine many of the ill effects flowing from a given industry. Consider the fact that (a) we do not know many of the short-term or long-term effects of industrial activity; and (b) we do not know many of the synergistic effects of industrial activity. If we do not know these effects how can we begin to calculate the costs associated with them and assign them to the appropriate source? Welfare economists have failed to follow through with the tools and techniques necessary to determine the monetary costs of social damage and, instead, assume that these costs can be assessed, and go on to argue for or against the application of this approach in the form of prohibition, government regulations, or effluent and damage charges.

The second difficulty is one of a judgmental or ethical nature. In the

event that the nature of the damage flowing from industry were deter-
minable, what costs would be assigned to loss of life, shortened life span,
disablement, psychological trauma, etc.? Setting these values is not a
question of science (economics or otherwise) but a question of what is
possible within a given political economy. If the values are set at a low
level, consumers may only have to pay a bit more for the privilege of
using products associated with human suffering. If the values are set at a
high level, many industries will be forced to close down as consumers
may no longer be either willing or able to afford to pay the full cost
associated with many products. In other words, even if the social cost
approach is possible, the values which are arbitrarily set will determine
whether the approach is preventative or merely compensatory. As with
many of the experts, the economists often substitute 'science' for reality
in an attempt to abstract themselves from the moral questions that flow
from their theories.

CONCLUSION

In this article I have been critical of the diverse professionals who have
shaped the present field of occupational health and safety. Generally,
the various disciplines have taken a compartmentalized approach and
have failed to keep pace with changes in the nature of work and the
workplace. My point of view may appear to be rather negative – particularly as far as the 'experts' are concerned. My criticism is merely an
attempt to define the problem; I feel that by understanding these negative aspects we can begin to approximate a positive solution.

It should be clear also that workers need more scientific research and
professional input in order to minimize the potential harm prevalent in
our workplaces, although worker reliance upon the experts ought not to
be a substitute for their own involvement in workplace health and safety.
We cannot ignore the contributions made by scientists, physicians,
engineers, and social scientists towards the elimination of dirty jobs and
the reduction of hazards. Practice is blind without science.

Too often society asks too much of its experts. We expect them to
provide the answers to every problem. Too often workers abdicate their
responsibility for protecting themselves and their fellow workers from
daily work hazards. We need the experts; they have the technical
answers to the technical questions. But we have fallen into the habit of
depending on them for too much – and this is as unfair to them as it is to
workers.

Too often, also, the conventional wisdom and its accompanying myths, such as the careless or accident-prone worker, have caused us to ignore the real causes of damage: working conditions. Clearly, work and the workplace are fundamental hubs around which human activity and behaviour revolve and evolve. There is a need for an understanding of socio-technical interaction and a need to apply such understanding in a manner that will enable workers to act, react, and interact in a manner consistent with their own health and safety. Worker involvement, in my opinion, is the key.

NOTES

1 Consider how this parallels the introduction of higher environmental standards. Plant shut-downs and moving the plants to locations with 'feasible standards' is often threatened, and sometimes even done, by industrial concerns.

2 A worker who was asked why he knowingly worked in an unsafe trench (it was unshored and about fifteen feet deep) told me: 'I don't dig ditches because I want to, I dig them because I have to.'

3 See, for example, G. Bruce Doern, *Regulatory Processes and Jurisdictional Issues in the Regulation of Hazardous Products in Canada*, Science Council of Canada Background Study no 41 (Ottawa 1977), 155: '... regulation of ... occupational ... hazards should begin in the workplace ... regulatory reform ... must begin there too. Labour and management are at the core of the system, and effective regulation depends on their joint co-operation based, however, on an institutional background in which both parties possess the necessary political, economic and legal "carrots and sticks" ... it is essential that a regulatory and legal policy based largely ... on the main features of the recently adopted Saskatchewan ... legislation be adopted both provincially and federally'; 155–6: 'Occupational health issues are in one sense part of the second historical phase of the reform of industrial relations. The first phase dealt with traditional economic needs and the right to bargain collectively. The second and current phase is concentrated on concern for overall industrial democracy, including economic health rights and how to take them out of the raw bargaining environment in which they have been historically (but regrettably) lodged.'

4 Occupational health and safety legislation in Canada and the United States has led to the development of an asbestos industry in Mexico, where no such standards exist. Also, some countries openly solicit foreign investment by guaranteeing that no health and safety standards will be introduced. In this light, many problems are merely being exported to underdeveloped countries as opposed to being eliminated.

5 This approach results in the use of pseudo-preventative techniques which are inadequate or unrealistic. The contents of an accident report from a large mine in the Maritimes illustrate this point well. In one instance an individual was 'walking at an excessive rate of speed' in the wash house when he slipped, fell, and injured himself. Under the column entitled 'Preventative Measures Taken,' it was stated that the worker '... was instructed not to walk at an excessive rate of speed.' In a similar case, a workman who had strained his back while attempting to push a cart with stiff wheel bearings was subsequently instructed in the proper method of attempting to push a cart with stiff wheel bearings.

6 For example, one Saskatchewan firm has a policy of letting employees go who upon

being tested are found to have an excessive concentration of lead in their bloodstream. Although workers are exposed to lead in the workplace, the implication is that excessive concentrations of lead in their bodies is their own fault. A similar example of 'blaming the victim' can be found in the safety rule book for the railroad industry which states that all employees are to work safely. Therefore, if an employee is involved in an accident he has violated this regulation and can be penalized.

7 See B.L. Hutchins and A.L. Harrison, *A History of Factory Legislation* (Westminster 1903), 13

8 See W.R. Lee, 'An Anatomy of Occupational Medicine,' *British Journal of Industrial Medicine*, 30 (1973), 112

9 A small town or company may have an advantage in this respect as the doctor will usually examine most of the workers in a particular industrial firm. For instance, a doctor in Charleston, West Virginia, observed that eight of his patients, all of whom worked with polyvinyl chloride (PVC), had a very rare but similar type of cancer and in this manner discovered the relationship between the two.

10 In a similar light, labour has expressed concern regarding the medical reports that go to most Compensation Boards: workers who are filing for a claim are not allowed to see their own reports, whether it be one made by their own physician, a board doctor, or any of the doctors on the special appeals committee.

11 Health Policy Advisory Center, *Health/PAC Bulletin*, no 44 (Sept. 1972)

12 United States, Department of Health, Education, and Welfare, 'Environmental Health Problems.'

13 Dr Irving Selikoff, director of the Mount Sinai Environmental Sciences Laboratory in New York City and perhaps the leading authority on asbestos epidemiology today, admits that he still does not fully understand the etiological and pathological characteristics of asbestos fibres. However, he is the first to admit that asbestos poses a definite health hazard.

14 These barriers, in the case of a negligence suit, were (*a*) the burden of proof was on the injured party; (*b*) 'the doctrine of common employment' stated that the victim of a negligent act of a fellow worker could not hold the employer liable (the rationale being that workmen implicitly contracted upon themselves, as part of the risk incidental to their calling, the possible negligence of their fellow servants); (*c*) the notion of 'contributory negligence' made the recovery of damages difficult if it could be shown that the injured party had contributed to the chain of events that led to the accident.

15 By way of example, the United States judicial system has defined a 'reasonable man' as one who 'never relaxes his vigilance under the influence of monotony, fatigue or habituation to danger, never permits his attention to be diverted, even for a moment, from the perils which surround him, never forgets a hazardous situation which he has once observed, and never ceases to be alert for new sources of danger.' This notion has obviously been carried over to describe the 'reasonable worker' as well.

16 Heinrich, *Industrial Accident Prevention*, 4th ed. (New York 1959)

17 It has been suggested that 'ergonomics,' a relatively new science that attempts to understand the nature of man-machine relationships, offers some hope in this area. However, it is clear that the ends towards which this science is applied will largely determine its usefulness in the area of occupational health and safety.

18 'Industrial Hygiene: Definition, Scope, Function, and Organization,' *AIHA Journal*, 20 (1959), 428

19 For example, a case was cited where the use of acid in pickling tanks caused rapid deterioration to the building. A hygienist solved the problem and 'the result was an

extraordinary reduction in the cost of maintaining the roof and wall structure.' *Ibid.*, 24 (1963), 206. No mention was made of lessening the danger to workers exposed to the fumes.

20 A.D. Brandt, 'Industrial Hygiene: The Broadening Horizon,' Cummings Memorial Lecture, *ibid.*, 207

21 Even when adjustments are necessary, they may not be made. For example, the TLV for sulphur dioxide in industry is currently 5 parts per million (ppm) – 50 times as high as the US Environmental Protection Administration warning level of .1 ppm. Workers are clearly being afforded less protection than the general public.

22 This apparently paradoxical statement should be explained. While it is obvious why industry should argue for weaker standards, it is less obvious why they should argue for stronger ones. The latter usually occurs when an industrial firm develops a new process which, as a side-effect, reduces the level of concentration of certain hazardous substances in the workplace. By pushing for the introduction of this new lower level, the firm hopes to gain a competitive advantage over its competition which must incur additional expenses to meet these lower levels and slow down production to re-tool. Also, if the firm is lucky its competition may even buy the technical details of the new process from it.

23 H.F. Smyth, Jr., 'Industrial Hygiene in Retrospect and Prospect – Toxicological Aspects,' *AIHA Journal*, 24 (1963), 222

24 See Karl Menninger, 'Purposive Accidents as an Expression of Self-destructive Tendencies,' *International Journal of Psycho-Analysis*, 17 (Jan. 1936), 6–16

25 Kerr, 'Accident Proneness in Factory Departments,' *Journal of Applied Psychology*, 34 (1950), 167

26 Greenwood and Wood and Greenwood and Yule did the original studies which, due to unclear thinking and eagerness on the part of others, were interpreted to indicate the existence of 'accident'prone' individuals. M. Greenwood and H.M. Wood, 'A Report on the Incidence of Industrial Accidents upon Individuals with Special Reference to Multiple Accidents,' British Industrial Fatigue Research Board, no 4 (1919); Greenwood and C.V. Yule, 'An Inquiry into the Nature of Frequency Distributions ... of Repeated Accidents,' *Journal of the Royal Statistical Society*, 83 (1920), 255

27 The best one may be the article done by A.G. Arbous and J.E. Kerrich, 'Accident Statistics and the Concept of Accident-Proneness,' *Biometrics*, 7, no 4 (Dec. 1951), 340. They drove home the point that the early studies had assumed that all the workers studied were exposed to identical social and technical risk and, therefore, that their involvement in accidents could be attributed to *personal* factors alone.

28 Ralph Nader has suggested that we begin to talk about 'corporate cancers.' He contends that if it is true that 80 per cent of the cancers are environmentally introduced, then cancers can be attributed largely to the production processes and products that flow from industry – undeniably a major source of environmental pollution. See 'Nader Calls Cancers a "Corporate" Disease,' *Labor Occupational Health Program Monitor* (Institute of Industrial Relations, University of California Center for Labor Research and Education, Berkeley), 3, no 2 (Feb. 1976), 7

29 Along this line, Terrance Ison, former chairman of the BC Workers' Compensation Board, levelled a monetary fine of $28,426.85 per month against Cominco for its failure to reduce the level of smoke, dust, etc., associated with one of its smelting plants. The fine was to continue until Cominco reduced the contamination level. The notion behind this was that, with the cost of compliance to regulations by industry usually being fairly inexpensive, industry has little incentive to comply if non-compliance is cost-free.

6
Occupational health policy in Canada

Robert Paehlke

There has been rarely a week in the past two or three years when some major Canadian newspaper did not contain at least one story related to occupational health. We have read of silicosis among Elliot Lake, Ontario, uranium miners, a variety of lung ailments thought to be more prevalent among several categories of steelworkers, generally increasing industrial deafness, black lung in Maritimes coal miners, native fishermen and their families in northern Quebec and Ontario in danger of mercury poisoning, risks associated with grain dust in Saskatchewan wheat elevators, and even radiation exposure for Port Hope, Ontario, school children. It seems as if, suddenly, there is almost no form of work that is not threatening to people's health.

The stories seem to follow a common pattern. Individual occupational health cases are raised by union officials, hospital or university-based researchers, an opposition politician, the news media, or some combination of the above. The cases are rarely put before the public by corporations, the parties in power, or the responsible branches of the civil service. On the contrary, these institutions and agencies generally seem to see their role as one to be played with caution, denial, and delay.

One reason for this posture of denial and delay is that it has been known for some time, largely privately, that many occupations involve severe health risks. If this were not the case, the officials, upon learning of the possible danger, would be as shocked and saddened as the public, and would proceed, carefully, to take action. For reasons that will be outlined shortly, it is difficult for any one actor – be it union, company, government, or independent professional – to carry through in ways that are equitable and satisfactory to all concerned. But it is nevertheless an indictment to have known or even suspected that there were dangers and not to have acted. Being responsible in any way for occupational health is not an enviable position. We can learn something of how difficult a matter we are dealing with by looking briefly at the history of one of the least ambiguous and most pervasive of workplace hazards: asbestos.

ASBESTOS

The truth is that asbestos is a deadly poison – a fact that has been known for centuries. As Jeanne Stellman and Susan M. Daum put it: 'Asbestos is a fibrous mineral that, like cotton, can be made into thread and cloth. Unlike cotton fibers, its fibers are as strong as piano strings. It is virtually indestructible – heatproof, fireproof, and resistant to most chemicals.' Its structure is made up of innumerable tiny fibrils, one million to the inch: more than a thousand times finer than a human hair. These fibrils are so tiny that they are easily suspended in ambient air and when inhaled bypass the protective hairs and mucus of the nasal passages and throat. They easily enter the lungs and 'it is estimated that in an 8-hour day, at the legal limit of 5 fibers per cubic centimeter of air, a person breathes 15 million fibers. Many of these remain trapped inside the lungs. Once they are inside the body they remain as indestructible as they were outside it. An asbestos fiber you breathe when you are 18 years old may stay in your lungs until you die.'[1]

The human body surrounds and traps that which it cannot destroy or eliminate. But inside the lungs scar tissue builds up around each fibre of asbestos and, in cases of major exposure, scar tissue increasingly takes the place of the delicate and critical air-sac walls of the lungs. The capacity and efficiency of the lungs is reduced accordingly. The central life-functions of the body are endangered. There is shortness of breath, a persistent cough, lessened efficiency of the heart, low-oxygen blood going to all the organs of the body including the brain, and, generally, a condition in which there is greater vulnerability to lung infection and to many other ailments. This is asbestosis. And after twenty or thirty years of coping with this reduced capacity, many persons (it is impossible at present to say in advance *which* persons) become more likely to contract a variety of cancers, especially lung cancer.[2]

Lung cancer deaths are six times more common among asbestos-insulation workers than among cigarette smokers. Of course, smoking increases the danger to these workers. Cancer of the stomach, colon, and rectum are three times more common among asbestos-insulation workers than among the population as a whole. And finally, meso-thelioma, cancer of the lining of the chest cavity, occurs only among people exposed to asbestos.[3]

It has been known that asbestos is dangerous since the time of the Romans. In modern times the specific disease asbestosis was first reported by a London (England) physician in 1900. In 1918 US and Canadian insurance companies stopped selling insurance policies to asbestos workers. In 1924 Dr W.E. Cooke reported an asbestos-linked death in the *British Medical Journal* and in the 1920s there were eleven such studies published in Britain. And by 1935 twenty-eight asbestosis cases had been reported in Great Britain and the United States. In 1935, 126 randomly chosen asbestos workers in the United States and Canada – most working for the Johns-Manville Company – were examined: 67 were diagnosed as having asbestosis, 39 were classified as 'doubtful,' and only 20 were seen to be completely free of any sign of the disease.[4] In the late thirties and early forties there were a series of cases reported indicating a possible link between asbestosis and lung cancer.[5] These were largely individual autopsy cases in which victims with a work record associated with asbestos were found to have had both asbestosis and lung cancer.

The form that the evidence took by the early 1940s was, of course, far from conclusive proof that asbestos exposure caused lung cancer. But let us take a moment to step back from the history reported here to consider its meaning for contemporary occupational health policy. In a sense, what we are dealing with here is the meaning of the word 'evidence.' I am going to argue throughout this essay that what is needed in the area of occupational health policy is a change in three things: (1) the acceptable operational meaning of the word 'evidence,' (2) the effectiveness of communication of newly found data from which conclusions can be drawn ('evidence assembled'), and (3) a shift in social attitudes so that the prevention of a hideous death (let alone thousands upon thousands of such deaths) from lung cancer is a higher social priority than inexpensive industrial pipes and household insulation. I believe that the need for the first two changes is demonstrable through an analytic study of the history of the asbestos industry and a human consideration of how it would *feel* to die of asbestosis and lung cancer.

The critical question to ask about the history of asbestos is *when* might one reasonably expect that a humane and intelligent government might have intervened on behalf of asbestos workers. On what grounds? In what ways? These questions are important, not to pin specific historical guilt, but to guide us in the present in a consideration of what to do about the very large number of substances used in industry today which may or may not be dangerous. Should we wait until we are relatively certain: until full epidemiological studies have been done, as they were done in the case of asbestos by Irving J. Selikoff and his associates at Mt Sinai Medical Center in New York using the welfare and retirement records of the asbestos insulators' union? Or we could act more decisively on indicative evidence, and limit the development of an industry until such time as either (1) exposures are foreclosed or (2) the substance is granted a clean bill of health by *independent* researchers, or perhaps even researchers hired by the *employees* of a given industry.

What might have happened, in the case of asbestos, had actions of that order been taken as soon as indicative evidence was available? The asbestos industry would clearly have been prevented from expanding immediately after the Second World War (wars are difficult times to curb industrial development). We would have waited until we were certain a proper level of safety was assured and it might have taken decades to learn how to handle asbestos in ways that allowed its workers to be as healthy as the population as a whole. This, of course, was not what was done.

The asbestos industry was tiny in the 1920s when the first modern links with asbestosis were being established, small in the 1930s when the first hints of cancer links were found, and booming during the Second World War (especially in the shipbuilding industry). Indicative evidence of danger had been available since the late 1930s. In 1951 the US Bureau of Mines reported: 'Production of asbestos in Canada, our principal source, surpassed all previous records by a wide margin in 1951. Nevertheless, the demand for asbestos has increased so sharply that most grades were in short supply.' Further, 'As in 1950, imports and apparent consumption exceeded all previous records. Imports from Canada again attained an all time high ... Acquisition of strategic grades for the National Stockpile continued to be difficult because industrial demand, of which a large percentage was directly or indirectly for the defense program, was quite urgent. Prices for Canadian asbestos were 10 to 25 percent higher in 1951 than in 1950.'[6]

Virtually no action was taken then in North America which affected the operation of the relevant industries in any way. It was another twelve years before Dr Selikoff and his associates were able to publish any truly conclusive results. Reviewing the history of their work, they wrote: 'In 1963, information became available which indicated that asbestos insulation workers employed in the construction industry were subject to a significant cancer hazard associated with their work (Selikoff *et al.*, 1964). Studying the mortality experience of the 632 members on the rolls of the New York–New Jersey branches of the insulation workers' union on 1 January 1943, for the 20-year period ending 31 December 1962, it was found that death from cancer was almost three times as frequent as expected. The death rate from cancer of the bronchus and pleura was 6.8 times as high as that for the general US white male population, both age and date being taken into consideration; and cancer of stomach, colon and rectum was found to be three times as common. Attention was called to the occurrence of pleural and peritoneal mesotheliomas (Selikoff *et al.*, 1965a).' The same year the US Bureau of Mines reported simply that 'Total U.s. sales (in 1964) were 52 percent higher in volume and 59 percent higher in value than in 1963. The Nation ranked sixth among world producers. Canada continued to be the leading producer with 40 percent of the total output.'[7]

Clearly, at least indicative evidence that asbestos was dangerous *preceded* the massive growth of the industry. By the time firm evidence was in, many workers had been stricken with asbestosis and some had died of lung cancer. Worse still, the industry is now so important that even if it turns out that asbestos cannot be handled safely under any circumstances, it is unlikely that, short of a major transformation of social values and social structures, we would deem it possible to do without asbestos, or some substitute likely to be equally as dangerous. But it is important to realize *for the future* that slowing the growth of the asbestos industry in 1945 would have been a *far* less difficult political task involving *far* less social and economic disruption. This understanding should influence how we proceed with the thousands of other potentially dangerous substances present in the modern workplace.

Throughout the 1950s dozens of new uses were found for asbestos. By 1964 it was used in the construction of office buildings, schools, factories, and housing; in asbestos cement pipes; flat or corrugated-sheet roofing and siding shingles and wallboards; safety clothing; clutch facings and brake linings in cars, trucks, and buses; floor tile; and as a filtering material. Thus, in the post-war period there was a slow and sporadic evolution of evidence of the dangers of asbestos and, simultaneously, a massive growth in the industry: growth in the volume of asbestos produced, in its dollar value, in the range and geographical dispersion of the industries in which it was used, in the number of workers handling it, and in the number of people exposed to it in the environment generally.[8] World production rose from about a million tons in 1947 to 3.6 million in 1965 to 5.1 million in 1975. The Canadian share of total world production has gradually fallen from 66 to 33 per cent in this period (1947–75).[9]

What about on-the-job protection in asbestos-using industries? The

record is so blatantly inadequate in Canada that a few items will suffice:
'In 1931, the British government made asbestosis a compensable disease
under its workmen's compensation laws, and some preventative measures were taken to limit worker exposure to asbestos fibers. In the U.S.,
where both workmen's compensation and occupational health programs
were under State jurisdiction, similar actions were not taken until the
1960's.'[10]

But workmen's compensation for asbestosis and lung cancer victims
was still under contestation in Ontario in 1975 – close to two thousand
years after the first indications of a relationship between asbestos and
the disease asbestosis,[11] forty-four years after general medical acceptance of the link in Britain, and a full decade after the publication of the
conclusive findings of Selikoff. Later in 1975 Romeo LeBlanc, acting
federal minister of the environment, announced regulations to limit
asbestos emissions from asbestos mining and milling operations: 'the
proposed regulations would limit the concentration of asbestos fibres
emitted to the ambient air to 2 fibres per cubic centimetre from crushing,
drying or milling operations and from dry rock storage ... The proposed
regulations are intended to protect the public's health, said Mr. Le-
Blanc. On the advice of the Department of National Health and Welfare,
Environment Canada is adopting a prudent course about major sources
of man-made asbestos emissions. Accordingly, the regulations cover
sources at mines and mills, including crushing, drying, milling and dry
rock storage. Further regulations are being developed for the other
sources in asbestos mines and mills, such as drilling operations, tailings
handling and tailings disposal ... Periodic emission testing as well as
record keeping for malfunctions and breakdowns are required also.
Estimated Canada-wide emissions of particulates containing asbestos
fibres from this industry were 80,000 tons in 1970. The application of the
proposed regulations would drastically reduce the emissions of asbestos
fibres to the ambient air.'[12] One wonders if the use of the word 'prudent'
is intentionally disingenuous given that prior to the announcement of
regulations scientific doubts had already been expressed regarding the
usefulness of the two fibres per cubic centimetre standard.[13]

The history of the asbestos industry and asbestosis suggests many
questions but especially this critical one: why was the necessary extensive research not carried out as soon as the first clues of danger started
coming in rather than in the 1960s? Obviously, the answer involves a
large number of factors including the following:

1 / In the early part of this century the number of workers involved was

very small and at the time most miners worked sixty or more hours per week under extremely dangerous conditions for a wage on which they could barely keep themselves and their families alive. They often were not unionized and, if they struck, they were subject to physical attack and arrest. In that context asbestosis was, frighteningly, a lesser trouble.[14]

2 / The tradition of scientific caution and neutrality and the class position of the scientists and doctors precluded their doing other than publishing their results in scientific journals read only by other scientists.[15] And scientists, by themselves, could do nothing about the situation.

3 / The corporations involved sought systematically to sponsor research that, on the whole, tended to delay relative certainty regarding the dangers: report summaries that minimized the import of their own findings,[16] facile reminders that correlation does not demonstrate causation,[17] straightforward pseudo-scientific whitewashes on a massively expensive scale,[18] careful studies indicating that some forms of asbestos fibre were markedly more dangerous than others.[19] Particularly common were studies introducing additional variables into the equation – for example, trace metal impurities in the asbestos[20] or the polyethylene storage bags in which the asbestos was packaged.[21] Each of these studies tended to compound the questions involved and, in a political context where *any* excuse will delay action, to buy time within which expansion of the industry could continue. I am not disputing the desirability of considering many of the matters which these studies raised. But it is a central argument of this paper that remedial action must be taken first, on mere indicative evidence, when human lives are involved. After people's health is protected, the full range of research questions can and should be opened.

4 / Researchers and potential researchers had no access to the health records of asbestos company employees. They were thereby unable, even after statistical and other techniques were fully developed and had been used on innumerable previous occasions, to carry out full epidemiological tests on long-term employees or ex-employees of firms producing or using asbestos. Finally, in 1955 R. Doll, a member of the Medical Research Council of England, was able to gain access to the data from records kept by the British government under the asbestos legislation of 1931 and do a plant-wide study. And in the early 1960s Dr. Selikoff and his associates obtained the records of the Insulation Workers' Union and did the first independent industry-wide studies.

5 / Another factor to be considered is the attitudes of the workers and

their unions in the industry. Few studies have been done that are useful in this regard, but it has generally been shown that clear and concrete demands or forceful actions on behalf of worker health were not made by the asbestos workers as a whole. Some reasons for this will be discussed later.

It seems to me that our consideration of the history of asbestos raises at least the following matters of concern: (1) Are we fully confident about the scientific neutrality of our scientific research processes (for example, are funding sources neutral)? (2) Are we fully confident about the effectiveness of the present means of scientific communication, especially between science and government? (3) Are we fully confident that every potentially dangerous substance is being thoroughly tested? (4) Given the possibility of long time-lags prior to effect on human health, should we wait twenty to thirty years before introducing any new substance into the workplace? (5) How much evidence of what kind should be the basis for decisions to interfere in the growth of an industry by the imposition of more stringent use regulations (we must remember that lives have been endangered by the long delays in asbestos-asbestosis-cancer research)? (6) Can we possibly set maximum exposure standards without, in effect, using human guinea pigs (the best hope lies in really effective control techniques which eliminate exposures almost completely)? (7) Should science more directly serve the people most affected by its findings or by its non-findings (for example, should the unions themselves hire scientists, medical technicians, doctors, and/or offer research contracts)?

These questions become especially pertinent when one considers three further characteristics of the contemporary industrial health scene. First, thousands of new substances are being introduced each year. Stellman and Daum put forward the figure of 6000–12,000 *toxic* industrial chemicals in *common* use and suggest that as many as 3000 *new* chemicals are introduced into industry every year. They further note that standards are developed for only about 100 new chemicals per year.[22] Obviously very few, if any, of the substances in question will prove to be as deadly as asbestos; but some may well be as bad, many are doubtless very dangerous, and unforeseen lethal combinations may occur. We must consider what procedures can be developed to make sure that mankind does not have to relive our experience with asbestos over and over again.

Second, we must realize that exposure to working conditions that can endanger health is the norm in North America, not the exception. In

1968 a past US surgeon-general reported before a congressional committee that United States Public Health Service studies showed that 65 per cent of industrial workers were exposed to toxic materials or harmful physical conditions such as excessive noise or vibrations.[23] There is no reason to suspect that this percentage has declined since 1968 or that it is any lower in Canada. That is to say, modern industrial workplaces in general are filled with an almost incalculable number of dangerous substances and processes.

And third, one can reasonably suspect that many of these hazards affect people in ways that are difficult to detect. For example, effects are always statistical – that is, they affect people unevenly; some are utterly unaffected through a lifetime while others are severely affected by a short exposure. Further, most of the substances have delayed effects, which show up often only after those exposed have changed jobs, moved to another community or province, or even after the offending substance is no longer used. In other words, studies of health impacts must be conducted for all workers, present and past, and over long periods of time. This, clearly, is a large undertaking, fraught with difficulties – including matters related to individual privacy – but even a brief look at the nature and range of workplace hazards will, I believe, support the call for a task of this magnitude, facilitated as it now is in Canada by the centralization of health records in each province.

A BRIEF CATALOGUE OF WORKPLACE HEALTH DANGERS

In one section of an essay one cannot, of course, include more than a short list of health hazards in the workplace. A fuller compendium is available in *Work Is Dangerous to Your Health* by Stellman and Daum. In this section I have followed their over-all categorization of hazards, but have included only some of the most dangerous, emphasizing those particularly relevant to Canadian workplaces. I do not include at any length consideration of workplace safety or accidents (including therein the whole field of ergonomics) as there is not sufficient space here. I have also excluded health risks from rotating work shifts and direct dangers of bacterial or viral infection on the job for the same reason.

Metals and minerals
Beryllium, a metal used to harden various alloys, is deadly, especially as a dust in ore-processing plants and in smelters. Beryllium lung disease,

known as berylliosis, can be induced by either brief intensive exposures or prolonged low-level exposures. The disease is frequently fatal, and even those who recover may have sustained permanent lung damage. There is a disabling chronic stage which may only develop after years of exposure and mild symptomatic effects. Beryllium ore is not mined in Canada though the metal is used quite widely in industry.

Cadmium is more common in Canada, and more deadly; it is most dangerous as a fume resulting from the welding or firing of cadmium-coated metals. In 1973 more than four million pounds of cadmium valued at more than $15 million were mined in Canada. Poisoning can be acute, but the chronic effects are perhaps more threatening and include a severe emphysema wherein the cadmium disintegrates the air-sac walls of the lungs. Also affected are the kidneys, the bone marrow, and the sense of smell. And, as Stellman and Daum put it: 'poisoning may become evident long after exposure has ceased, and may progressively worsen without further exposure ... One scientist has stated that it is impossible to predict a "safe" level because the cause-and-effect relationships between exposure and illness vary so much from case to case.'[24] These two attributes are not uncommon among the more hazardous of industrially used substances.

Lead poisoning and mercury poisoning are classic cumulative poisoning diseases. The body stores them for years and responds negatively after the slow accumulation of toxic quantities. Inorganic lead is stored in the bones, organic lead compounds in the brain, and mercury and its compounds in the kidneys. Both substances are widely distributed in Canadian workplaces and in the environment generally. Mercury builds up in food chains and is a hazard to those who subsist on fish. Lead from the paint in older buildings is a hazard to children. And lead has been distributed everywhere in Canada by the automobile and by hunters. Since both dangers are cumulative and almost every Canadian has had some exposure, those whose workplace environments contain additional amounts are that much more endangered. Lead and mercury poisoning result in damage to the brain and the central nervous system. Short of crippling brain damage and death, there are irritability, dizziness, speech impediments, loss of appetite, headache, personality disorders, weakness, and loss of co-ordination. The expression 'mad as a hatter' derives from the brain damage induced in hat workers who, at one time, were generally exposed to high levels of mercury fumes. Small amounts (circa $1 million) of mercury were produced in Canada in 1973; 753 million pounds of lead was mined, valued at $122 million.

Critical to thousands of Canadians is the fact that there is some evidence that nickel dust is related to cancers of the respiratory tract and nasal sinuses. There has been considerable evidence that cancer deaths were notably high in certain facilities operated by the International Nickel Corporation near Sudbury, Ontario. Stellman and Daum summarize: 'Contact with nickel carbonyl fumes, and with other forms of nickel that produce dust and fumes, should be very limited, and adequate ventilation and other engineering controls should be provided. Skin contact should be avoided by carefully planned industrial procedures, and workers should be provided with protective clothing and facilities and time for frequent washing.'[25] In 1973 the nickel industry in Canada produced 550 million pounds of nickel valued at $800 million. Canada is the world's leading producer of nickel, and nickel ranks tenth among *all* Canadian exports. The environmental record of this industry is generally among the worst in Canada.[26] So far as I am aware, there has not to date been an industry-wide epidemiological study conducted in Canada for present and former nickel miners and smelter workers; a large study is now under way at McMaster University medical school.

Common inorganic gases
Sulphur dioxide is produced in staggering quantities in Canada. The facilities of the International Nickel Company in Sudbury alone produce about 3600 tons per day.[27] Most of it is spread a hundred miles and more in the prevailing wind pattern by the famed INCO high stack. In addition, sulphur dioxide is generally produced whenever coal is burned (though there are some low- and no-sulphur coals), when most metals are smelted, and will be produced in very large quantities in future tar sands processing plants. One can taste sulphur dioxide at less than one part per million, and smell it at 3 ppm. At 6 ppm it sharply irritates the nose and throat and at 20 ppm can result in eye inflammation. It is associated with chronic irritation of the nose and respiratory tract, changes in senses of taste and smell, and increased fatigue. Chronic nose and respiratory tract irritation may cause chronic bronchitis and emphysema. Little is known about the effects of long-term exposure to sulphur dioxide, but there are increasing grounds for suspecting that it is indeed dangerous.

Carbon monoxide is associated with most combustion: from furnaces and forges to internal combustion engines. Among those most commonly exposed are traffic policemen, those working in vehicle terminals and tunnels, those working on or near busily travelled streets, and those in many other warehouses and factories. Carbon monoxide acts chemically to replace oxygen in the blood; exposures first affect the brain. Symptoms include headache, weakness, dizziness, hearing loss, vision loss, nausea, and – at higher dosages – coma, suffocation, and death. It is particularly dangerous for those whose jobs require alertness: for example, taxicab drivers. Stellman and Daum report the following as well: 'Effects of long-term, low-dose exposures have been studied. Swedish workers exposed to concentrations that caused carboxyhemoglobin levels of 10 to 30 per cent showed such symptoms as headache, dizziness, decreased hearing, visual disturbances, personality changes, seizures, psychosis, palpitation of the heart associated with abnormal rhythms, loss of appetite, nausea and vomiting.'[28]

Hydrogen sulfide has the odour of rotten eggs. It is present in petroleum refining, a variety of chemical industries, and in the production of heavy water for nuclear power plants (a rapidly growing industry in Canada). Again, the effects of long-term, low-level concentrations are not fully understood.

Common dusts
Perhaps the classic case of occupation health dangers is that of exposure to coal dust. Thousands and thousands of coal miners around the world have died premature deaths as a result of black lung disease; hundreds of thousands have contracted the ailment. It was not until the 1968 West Virginia general strike of coal miners that an acceptable state compensation law was passed there. As Franklin Wallick reported: 'For more than twenty years the British had hard scientific proof that coal mining caused Black Lung ... Yet for those same twenty years our own u.s. medical experts ... would not assert as the British medical profession did, that coal miners got black lung from working in the mines. Thus for twenty years American coal miners suffered serious impairments to their health and were denied any compensation rights or any remedy for occupationally-caused impairment of health. This black-lung fight is a terrible indictment of American medicine.'[29] Compensation, of course, is not protection. The (us) Coal Mine Health and Safety Act of 1969 imposes air quality standards for coal mining. Relatively few mines have approached what might be considered safe levels for long enough for us to know what the effects are of relatively low exposures to coal dust over long periods of time.

Silicosis is a disease common to those who work in a wide variety of mines and construction sites. It results from the action of crystalline silicon dioxide dust on lung tissue. There is always shortness of breath –

at first when the victim is active, and with continued exposure even when at rest. The lungs become stiff, blood oxygen is low, and there is added work for the heart. There is increased susceptibility to tuberculosis and other lung infections and to heart disease. Chronic silicosis takes years to develop, usually arises years after the exposure period, and can become progressively worse without further exposures. The number of cases in North America in this century is clearly in the hundreds of thousands, at least.[30] For the uranium miners of Elliot Lake, it may be linked as well with lung cancer, though there is no clear evidence on this as yet. There is no certainty about a safe level of exposure.

Byssinosis is a lung disease of those exposed to cotton dust; Stellman and Daum report that 17,000 us cotton workers suffer from this disease. It is an allergic reaction in the small air tubes and involves chronic inflammation of the air tubes: chronic bronchitis. It can result in emphysema. In acute attacks it is similar to asthma: shortness of breath, dry cough, severe air hunger, chest tightness, and wheezing. The us federal government estimates that 12 to 30 per cent of cotton workers have the disease. Cotton dust is difficult to detect and difficult and expensive to control. As with asbestos and coal there is here a history of doubtful conclusions in industry-funded 'independent' research.[31]

A dust-induced disease of special concern to Canadian workers is farmer's lung, a result of work with organic dusts, especially grain dusts. It is common to grain elevator workers, ship and train loader workers, grain terminal workers, and others. The dangers have been known for centuries: 'The throat, lungs and eyes are keenly aware of serious damage; the throat is choked and dried up with dust, the pulmonary passages become coated with a crust formed by dust, and the result is a dry and obstinate cough; the eyes are much inflamed and watery; and almost all who make their living by sifting or measuring grain are short of breath and rarely reach old age' (Bernardino Ramazzini, *Disease of Workers*, published in 1713[32]). In 1971 the Grain Services Union in Saskatchewan was able to get workmen's compensation benefits for the effects of grain dust.

Organic chemicals
There are thousands of organic chemicals used in contemporary industry. Most have attained widespread usage since the Second World War and their long-term effects are generally unknown. However, acute effects of many are well-known: dermatitis, eye damage, seeming drunkenness, respiratory damage, headaches, fatigue, and so forth. Among

the more dangerous organic substances in the workplace are benzene (which has been linked with chromosome damage, birth defects, and leukemia), naphthalene (liver and kidney damage), phenols (damage to central nervous system and circulatory system), alcohols (dermatitis, eye damage, liver and kidney damage), ethers (dizziness, pneumonia, pulmonary edema), and glycols.

Recent findings have brought to light the extreme danger of working with vinyl chloride, an ingredient common in the manufacture of many forms of plastic. Thousands of workers have been exposed in plastics manufacturing plants, electrical wire and equipment plants (where it is used in wire insulation), plastic mouldings and fabricating plants, and elsewhere. For three years the B.F. Goodrich Company suppressed the information that vinyl chloride caused liver cancer in rats.[33] It is simply not known how many persons will eventually now contract cancer from exposures to vinyl chloride. There is a considerable time-delay in cancer development. In August 1975 B.F. Goodrich announced new processes for removing the deadly vinyl chloride monomer (vcm) from polyvinyl chloride (pvc) resin and claimed that 'the process produces resins with such low residual vcm level that processors and fabricators who use the resins probably will not have to establish regulated areas under the federal Occupational Health and Safety Administration (osha) standard covering worker exposure to vcm ... In addition, the process is expected to reduce vcm emissions from pvc plants to a level that will be in compliance with regulations which have been discussed by the Environmental Protection Agency (epa).'[34] If one assumes that the regulations discussed are indeed safe, one is left to draw the conclusion that the technology of safety was possible all along; its introduction was delayed – perhaps at a cost of human lives – because no independent agency tested the substance prior to its use in the workplace.

Another plastics industry family of chemicals, polychlorinated biphenyls (pcbs) have become a widespread environmental hazard – making many fish in the Great Lakes inedible. pcbs were known in the 1930s to cause a variety of skin disorders in workers, yet little was done to slow their increasingly wisespread use. Now they are being phased out of production by Monsanto, the only producer. As of January 1976, they were used only as high-efficiency insulators in enclosed high-voltage electric transformers.[35] Because we did not act decisively when workers' health indicated a danger, we now must live with widespread environmental contamination.

Carbon tetrachloride is widely used in industry and repeated low-level exposures or a single high-level exposure can cause severe liver and kidney damage; it is also a carcinogen. It is used as a cleaning agent, even as a household cleaner. Chloroform, the anaesthetic, is also a threat to kidneys. Both, it is feared, may be finding their way into public water supplies. Again the list of organic compounds known to be dangerous is seemingly endless. But how dangerous, at what concentrations, over what time period, for what length of exposures, in what combinations, is only rarely known.

Noise, heat, and other radiation
Hearing loss is one of the most common health problems brought about in the workplace. It is estimated that 17 million workers in the United States are regularly exposed to dangerous noise levels on the job.[36] There is no reason to expect that the Canadian record is relatively better. Many, many workplace settings are noisy – this includes those that are relatively less threatened by chemical dangers: for example, automobile assembly lines, airports, and construction sites. Noise compounds the hazards in such work locations as mines and smelters. Many workers suffer temporary threshold shift (TTS) wherein they temporarily cannot hear lower volumes; some later suffer permanent noise-induced threshold shift (PNITS) and loss of hearing range. But perhaps more importantly, high-level noise in the workplace has been shown to be associated with heart problems, circulation problems, balance, high blood pressure, heart discomfort, and digestive difficulties. And high noise levels greatly increase the likelihood of workplace accidents.

Heat in itself can be severely dangerous even if it is not sufficient to cause burning. It is suspected that the much higher cancer incidence among coke-oven workers may be as much a result of exposure to high heat as it is of inhalation of the gases and dusts released in the process. A US study found that those who work on top of coke ovens have a lung cancer rate ten times that which might be expected. There are more than a thousand such workers in Canada. Adequate research has not yet been done on the effects of long-term exposure to either heat or cold.[37]

Also present in many workplace settings is ultraviolet radiation (eye damage, skin cancer), infrared radiation (eye damage), and microwaves (genetic effects). Perhaps even more common are radioactive materials and X-rays. X-rays are used to check on welds, and various forms of radioactive materials are shipped, used, and stored in many settings. Most dangerous perhaps are radioactive gases and dusts. The US Public Health Service estimates that between 600 and 1100 of America's 6000 uranium miners will die of lung cancer.[38] Radioactivity exposures are associated with most forms of cancer and with genetic damage; again, most of these effects are delayed.

GENERALIZATIONS

Occupational health hazards are present across the entire spectrum of occupations and industries. This is especially the case when one includes such variables as stress, lack of exercise opportunities, noise and radiation exposures. There are, of course, occupational sectors which are generally more hazardous than others, notably mining, metals processing, and chemicals.

Many of the most dangerous forms of bodily response to hazardous substances, for example, cancer, emphysema, heart disease, or damage to the nervous system, can result from long-term exposures and may, as well, involve a long period of delay prior to symptoms being visible or pronounced. Delayed effects occur for a wide variety of ailments: silicosis, black lung, lung cancers in response or partial response to organic chemical exposures, asbestosis, most cases of work-related emphysema, most serious hearing loss, lead poisoning, and mercury poisoning.

There is, generally, wide variation in responses to many of the health hazards we have discussed. Some individuals can work for thirty years in an asbestos mine, smoke cigarettes through their lifetime, and live to be eighty years old. Those exposed are merely statistically much more likely than the average person to contract a certain set of diseases including lung cancer. Yet there have been other cases which have linked even brief household exposures of wives and children to the asbestos dust on the asbestos workers' clothes to mesothelioma.

There has been a pattern of slow response by industry and government to indicative evidence. In established industries such as coal mining, response in the United States, for example, was painfully slow and the delay involved many examples of medical and research complicity. Corporate research facilities or corporate-funded research teams have also on many occasions seemed intentionally to seek to cast doubt on the degree of danger. They often assume safety which they then seek to verify. Policy-makers rarely, if ever, have assumed that there is a risk until danger has been irrefutably proven.

Several extremely hazardous substances have gained general distribution in industry and thereby into the wider environment even after indicative evidence of danger was available. Scientists who suspected a danger were generally unable to make their suspicions part of public knowledge. Some of the blockages here include the absence of publications read by and written for both research scientists and the general public, including labour or labour union officials; the absence of cross-class associational contacts between these same groups; the inability of relatively independent research scientists to attain access to health records or sufficient research; and the seeming failure of findings and conclusions to cross the Atlantic Ocean.

There are thousands upon thousands of substances which may or may not be dangerous to health used in a variety of workplace settings. When combination with other substances is taken into account, there are literally millions of potentially hazardous situations of which we are nearly totally ignorant.

Initial responses are often masked: sinus trouble, headaches, stomach trouble, flu, cough, bronchitis, dermatitis (even acne) – all ailments one may well have contracted without exposure to anything hazardous. Causal links are thereby camouflaged.

Many workplace health hazards involve irreversible effects, incurable ailments, or general systematic weakening. Lead and mercury poisoning are not readily reversible even if intake is halted. There are very few who survive lung cancer. More generally, when lung, circulatory, heart, liver, or kidney function is impaired life expectancy is almost always lessened.

Exposures to hazards within a given workplace vary considerably. Precise work location is often critical. Those on night shifts may be less or more exposed. Different parts of a factory or mine have higher or lower gas or dust levels. Epidemiological tests of whole industries or occupations can fail to expose some particularly hazardous tasks. Furthermore, public health records in North America are generally of low quality and control groups are difficult to assemble. An industry as a whole may appear to be only a small percentage more dangerous than life in the society at large, while small groups within that industry are highly threatened.

Before turning to the policy implications of these generalizations I would like briefly to consider why it is that safe workplace health practices have taken so long to become a policy issue in North America. Obviously the question could be answered by saying that 'we' simply did not understand how the human body responded to various substances. We are only now beginning to understand: we are in an era of medical discovery akin to the first understanding of germs, sterilization, and immunization. The major causes of death remaining (heart diseases and cancers) are less often linked to micro-organisms, but are related to a wide variety of life habits: dietary patterns, cigarette smoking, food additives, environmental pollutants, lack of physical activity, and – to a far greater degree than we have realized until recently – workplace practices. It is in these areas that the future of medical research and practice lies. Medicine will clearly need to become more collectivized, more preventative, and more educative than it has been.

But what societies 'know' and do not 'know' needs explanation as well. That is, *why* did it take thousands of years for us to begin acting on suspicions of asbestos dangers? As I wrote the first draft of this paper (in February 1976), the CBC national news reported a mine in northern Ontario with asbestos dust in the processing plant so thick that it accumulates in piles several feet deep on the rafters of the building. There has been no lack of precise knowledge of the degree of danger for more than a decade. Clearly this is not a question of knowledge or the lack thereof; it is a question of values and of relative economic and political power. And it is these latter considerations which I submit are at the root of the original slowness to do sufficient research and to spread sufficiently widely what research knowledge had accumulated.

A full explanation of this phenomenon would require more space than is available here, but even a partial explanation must examine the following factors:

1 / the economic importance of the industries wherein the hazardous conditions exist to the political jurisdictions which have potential authority over those hazards;

2 / the integration of those industries within the economy as a whole (that is, for example, the dangers to other industries of precipitous changes);

3 / the socio-political power of the industries directly and indirectly affected: their influence over government through lobbying and other means, their influence on the media generally and on the autonomy and integrity of medical and scientific research; their influence, by their presence and seeming permanence, on social values;

4 / worker attitudes towards workplace health issues in the context of attitudes towards work, health, life, and politics; the absence of forceful occupational health education in the workplace;

5 / the severe difficulty which bureaucracies – government, union,

or private – have in overcoming inertia either internally or externally when the data on which they must operate is both difficult to obtain and incomplete or ambiguous.

Let us examine each of these matters briefly, for in studying the roots of past delays we learn something of the politics of occupational health. First, what is the significance to the Canadian economy of some of the relatively more hazardous of industries?[39] For example in 1976 mineral production in Canada was valued at $15,392,839,000. It was second only to agriculture among primary industries in net value of exports. Canada leads the world in the production of nickel, asbestos, platinum, and zinc; it is second in the world in the production of uranium, gold, cadmium, gypsum, and bismuth. When one considers as well the $139,105,000 worth of lead, $268,000,000 worth of coal, the noise levels of all mines, the gases and fumes of smelters, and the nearly ever-present silica and other dusts, one begins to grasp the problem: many of the most hazardous industries in Canada are critical to our economy as a whole.

They are also crucially important to the economy of the Western world as a whole. In 1970 we exported 96 per cent of the nickel we produced, 95 per cent of the asbestos, and 85 per cent of the copper.[40] These are not uncommon percentages. Any significant rise in the cost of our exports raises the likelihood that buyers will turn elsewhere for supply. Often, of course, the same corporation produces in a dozen or more different nations. Many operate in countries where labour is far cheaper and no one is about to raise workplace health issues. The corporations (when the economic system is working as it should) shift the largest share of their production to the least expensive location and their administrative officers in each nation therefore must seek accordingly to minimize costs in their locale. From their point of view this is in the best interest of the nation as a whole, the communities wherein they operate, and of the workers in their employ; and there is much truth in that view.

Reducing dust levels, for example, becomes generally increasingly more expensive as one approaches zero levels. As the levels are lowered the likelihood of health danger generally declines. How safe is worth how much to the most humane of those within the corporation? Who must decide? Is it not more convenient to believe that the danger is overblown, that smoking is just as unsafe, or driving a car, or being a stress-ridden executive? One might easily read too quickly the reports from one's company doctor (if it has one) and turn to other seemingly more pressing matters. And workers in some other country are easy to ignore, even if that country is a friendly northern neighbour. Easier still if it isn't.

But it is not only executives and corporate decision-makers for whom these industries are vital. The per capita value of mineral company production in Canada in 1972 was $293.53. Many, many towns and cities in northern Ontario and Quebec and elsewhere make no economic sense otherwise: no company would replace towns that extract what lies beneath them. There are 1716 mining industry firms in Canada. They employed 73,044 production workers in 1972; the payroll for those workers was $666,505,000 in that year. In addition, there were 23,509 production workers in smelting and refining with wages totalling $210,030,000. There are large clerical, sales, engineering, and management staffs supporting the workers. In brief, these industries are enormously important to many individuals; hundreds of communities exist only so long as the industries remain. Provincial and federal revenues and Canada's balance of payments ride on the relative competitiveness of these industries.[41] Cleaning up is expensive.

There are dozens of other industrial sectors wherein health hazards are common and many of them are even more globally mobile than the mineral industries; they can move more freely since their assets are not in the ground. Pulp and paper is Canada's leading industry (larger than motor vehicles) – it produced $5,703,192,000 worth of goods in 1974. Petroleum refining ranked fourth; and in the top forty industries were the following: iron and steel mills (6); smelting and refining (10); metal stamping and pressing (12); rubber products (15); plastics fabricating (19); industrial chemicals (29); miscellaneous industrial chemicals (35); and inorganic industrial chemicals (39). All involve a wide variety of health dangers which we have discussed. And the dangers are even more generalized than this.

For example, the Canadian electric wire and cable industry (36), which might be thought to be relatively clean, in 1972 purchased and used 128,599 pounds of asbestos. Further, they purchased and used 40 million pounds of vinyl resins, including polyvinyl chloride as the largest single item in that group. The pulp and paper industry in 1973 used 516,671 tons of liquid chlorine, produced largely in the chlor-alkali process, a heavy user of mercury and generally a hazardous industry.[42] The chlorine was valued at $31,685,000. The pulp and paper industry used as well 4276 tons of zinc dust, valued at $2,263,000. One cannot generalize about safety at point of use, but the zinc was mined and smelted. In any consideration of the politics of occupational health, industry has to be seen as tightly interconnected; regulating one industry affects many.

Lying beyond the economic power of industry is its enormous political power. Industry funds a wide variety of research – internal, university-based, and seemingly independent. For example, the asbestos industry as a whole has sponsored numerous studies of the effects of asbestos on human health for every study sponsored by a union or government. The enormous political power of industry is already well documented[43] and need not be further elaborated here.

Unless one were prepared to argue that industrial workers simply can never get their way under a capitalist system (or cannot get more than the now habitual share of productivity increases as salary raises), there is an issue of why labour has not been more successful in protecting worker health. On one level health can be seen as a problem qualitatively different from those which have been handled in the past by the North American labour movement as a whole. The first concern of the labour movement has been wages, second (since the advent of the 40-hour week) probably social insurance compensation for the aging, the sick, and widows, and third, grievance-handling over concrete and individualized conditions regarding supervisory attitudes, workplace comforts, and conditions of work. Occupational health issues, as we have seen, differ in several important ways from all of these other concerns. Critically, they are less tangible and immediate than most grievances: often there are no obvious victims who can do the complaining for themselves; most who know are too sick to raise the issue forcefully with the company or the union. The effect of a raise in pay, the institution of a dental health insurance plan, or the alteration of a work schedule or supervisor have immediate and obvious effects. The adjustment of a TLV may not even be perceptible and its effect will likely only be known statistically and, at that, not for twenty or more years. And studies of worker attitudes have shown that many do not expect to be at their present job for a very long period of time.[44] Often they are living on dreams of lottery wins or of going into business for themselves; whether these dreams are realistic or not, long-term issues are not matters of first concern. Earning enough to support one's family is.

In addition, workplace health issues are different from those which unions and employers are in the habit of considering in collective bargaining procedures. They often involve alterations in work routines, equipment design, materials use, supervisory authority (sometimes requiring *more* authority over disbelieving risk-takers), matters which are seen in many industries as 'management prerogatives.' With strenuous pressure by the unions the definition of 'management prerogatives' might be adjusted – but often they are maintained 'for a price' (wage increases). It has been quite a common procedure to include workplace health on the list of contract demands to be traded off in the bargaining process for higher priorities when management's back went up.[45]

Another consideration here is that workplace health issues require a range of technical expertise not common to most industrial unions (or to any other institution for that matter): a combination of medicine, industrial engineering, chemistry, and statistics. Union officials who come up through the ranks develop such skills as economics or industrial engineering or hire those who have found them, but very few indeed even have access to the range of skills needed to deal with occupational health issues. Furthermore the other technical matters are more in keeping with the actual workplace or home experiences of workers in the plant who often deal directly with, or know personally, engineers, accountants, or time-study experts. Workplace health issues involve skills that are less familiar to most people and are thereby a bit mysterious and difficult. Shop floor understanding and concern is often very low for these reasons. For the most part then, neither the union memberships nor the unions themselves are at present equipped to handle these issues.[46]

Another point which might be made here – though empirical documentation is not available – is that part of the attitude-set of North American industrial workers has centred around the 'virtues' of toughness, masculinity, and independence in the face of adversity. The auto industry and the tobacco industry to a great extent have fashioned their advertising product design and very existence on the role-conceptions of North American males. These same role-conceptions often make it a matter of pride that one treats hazards with a sense of abandon. Above all, one never shows or admits weakness, never indicates that the job is too tough. Before workers fully understand how dangerous silica dust, various industrial gases, or noise are they just endure the discomfort and pride themselves on their physical strength and inner worth. And in fact at point of first exposure – as we have seen – effects are often easily ignored or imperceptible. There are still endless examples of disbelief and indifference today, even with known killers like asbestos. In a recent issue of *Labour Canada*, a case of asbestos workers blowing rather than sucking the dust off hoppers with air hoses illustrates the indifference that can exist. Some of this indifference can be explained in part in terms of sexual role-typing, some equally well in terms of alienation and a need to appear 'cool.' But these are psychological explanations that are inevitably incomplete.

On the other hand, there is some evidence that health and safety issues are high on the list of worker concerns. As Wallick reported: 'Health and safety hazards rank as the number two complaint (to inadequate fringe benefits) of all American workers in a scale of nineteen sources of workers' discontent. This is a finding of a u.s. Department of Labour study on working conditions completed in 1971.' And yet this concern continues to be traded off for other sorts of gains. A reading of *Dying Hard* by Elliott Leyton gives some indication of the roots of the problem.[47] Those who went to work in the fluorspar mines of Newfoundland first did so in the 1930s and 1940s at a time when there were *no* other options. Fishing was destroyed by tidal waves and market conditions and working in the mines was the only way to avoid cold and hunger. The chance to work in utterly unregulated mines was welcomed, even celebrated. Dust levels were brought down around 1960 when it became clear that nearly everyone who worked in the mines had been sentenced to an early and painful death. And yet today workers continue to work in the mines – there is still no real choice in the area – and the level of hazard under the newer, presumptively safer conditions has yet to be determined with any precision. Most mines in Canada are located in poor, isolated communities where there are few other employment opportunities, unemployment is high, education levels low, and many have family homes that could not be replaced elsewhere where higher prices prevail in housing. Many are deeply in debt with mortgages and payments to banks or finance companies for cars or furniture. And many of the most hazardous occupations are those which go through regularized boom and bust cycles: workers are very hesitant to provoke slow-downs or closures. And to workers used to an unsafe daily working atmosphere, conditions dangerous to long-term health must often seem secondary.

It is one thing to indicate concern in an opinion survey and quite another to take appropriate action: to strike for a healthier work environment. It is not an easy thing to take action when generations have worked under even worse conditions and when the hazards are not even fully understood by scientists. Workers typically do not even know what it is they work with. The generic labelling of chemicals is not required in any jurisdiction in Canada. Canadian society as a whole has valued the lives of these people very little, and they often seem to show a corresponding lack of consideration for their own lives.

Lastly, let us consider the position of those in the civil service bureaucracies who are charged with setting health regulations for industry. They must act in defence of the relatively weaker position of those who work in industry (weaker whether the workers are unionized or not) and against the economic interests of industries whose executives generally must make decisions that maximize profits. The civil servants in this political context must find it difficult to initiate actions when the evidence of health danger is less than absolutely conclusive. But, as we have seen, there are often long lags before disease shows itself; there are many other possible sources of disease; hazardous substances have uneven effects on various human bodies; and there is considerable geographic and employment mobility of workers in the period between exposure and effect. Evidence of hazard is thereby always difficult to obtain and often less than fully conclusive.

The importance of the ambiguity of data is only understandable in this context. It is critical in any decision for concrete action: whether or not to set standards, how much funding to request for enforcement, how low to set standards, and so forth. Health administrators within industry and government would be expected to tend to look for irrefutable evidence of health damage. Is it not easier to suggest research towards ruling out yet another intervening variable before requiring actions that might be blocked at the political level? Is it not easier to adopt the TLV s of another political jurisdiction – a *concrete* request – than to argue for independently set levels within one's own jurisdiction? The levels in almost all cases are necessarily arbitrary: they balance a merely probable health risk against seeming economic impossibilities – often calling for actions demanded by no one. For example, we will not know what a safe level of exposure to vinyl chloride is until we set a level and study the effects over the next thirty years on all those who are exposed at those levels. What degree of risk demands the closure of an industry until changeovers can be made that totally eliminate the hazard? What degree of risk demands massive expenditures on protective devices? Stringent controls could lead to a hazardous level of unemployment as well, considering how many of our industries use dangerous substances and processes, and how little we know about the effect of so many of them.

Acting on the basis of firmly documented, quantitative evidence is a thoroughly entrenched habit of modern bureaucratic societies. It is perhaps even the hallmark of modernity itself. Our entire educational system is built on this precept; the very definition of scholarship itself seems in contemporary society to rest on this need. We, in effect, require figures demonstrating exactly how many workers must die, or have their hearing impaired, or become unemployable *before* we take any action, however inexpensive. There seems to be no number of lives that would require the closure of an industry of any size. It took the government of Canada forty-five years from the first suggestive evidence of danger to set a TLV for asbestos exposure, and it has not yet been firmly and uniformly enforced. Are we unable, one must ask, to allow 'likely lives' to tip the balance against 'certain dollars'?

The tone of this brief section is seemingly unscholarly and that is the point. It can be stated with certainty (of a statistical variety) that human lives will be lost if precise evidence is required in all future occupational health cases prior to action being taken. Unless we assume that there are no values higher than precision (or money), regardless of what else we do, we must qualitatively advance the level of action-relative-to-evidence. This is the general injunction, but what specific actions can be taken?

We must first realize just how enormous and difficult a research task we are facing. In the period of delay between exposure and effect many workers will have changed occupations, jobs, employers, residence, community, and doctors. Many will also have changed province or nation of residence and even dietary practices and personal habits. Many will have died of causes utterly unrelated to the health hazard – they will not have lived long enough to contract an occupationally induced disease. And they may have died and have had a faulty cause-of-death diagnosis. Or they may have died of a secondary effect of the health hazard to which they had been exposed which was aggravated though not caused by the hazard. An example is heart failure caused partially by obesity, tension, or mere 'aging' and partly by lung damage which added to heart burden.

These sorts of considerations lead one to the conclusion that we cannot do an adequate job of research without some greater centralization and standardization of medical records. Such an effort, of course, would be enormously helpful in the treatment of individual cases throughout life. It is the only way to cope with what seems to be an ever more mobile population and an ever more complex environment. People's medical records should follow them whenever they move: from on-the-job to off-the-job, from physician to physician, from province to province. They should be available to any physician who treats them, even in emergency situations. Persons whose employability is affected by their medical history must somehow be compensated.

Second, merely given the *number* of possible hazardous substances and the complexity of human response there is little hope at present of determining *which* substances are hazardous at what concentrations and frequency of exposure, and in what combinations, and which individuals are particularly susceptible and which are relatively less susceptible. We must massively up-grade the research effort and the sharing of data from nation to nation. A systematic international research effort is needed.

The number and complexity of potential hazards also suggests a need for recording all chemical exposures in the workplace which are either suspected to be unsafe or about which we are ignorant. This too is a massive undertaking, but I do not see how it can be avoided if we are ever to have safe workplace settings. If such data were incorporated into medical records, I suspect that its cost would be saved back again by society in terms of the *dollar* value in individual preventative medicine alone. That is, if it were a part of a central computerized medical records system it would allow for individuals who had had exposures to be called in more frequently at high-risk ages for tests and checks of various kinds. This could be done regardless of whether or not, say, an individual miner had changed provinces and opened a grocery store. The lessened need for treatment, and the decline in lost worktime, alone involve high savings. These are public savings with publicly supported medical care. Obviously, exposures need not necessarily be monitored individually at first. In the beginning they could be done on a workplace-by-workplace basis for a limited number of occupational sub-categories for each location. I am convinced that sooner or later such forms of medical and occupational record-keeping will have to be instituted. I see no reason not to begin studies (including even cost-benefit) on such schemes immediately.

Once such a double system of record-keeping was refined (and expanded perhaps to include data on personal health habits such as smoking, diet, and exercise habits, and pollutant levels in residential locations) a limited number of researchers could monitor all workplace practices on an on-going basis. One developed country could serve the world with advance warnings. Or the responsibility could be divided among several nations, each taking charge of the research for its own leading industrial sectors.

It would be a useful beginning even if all that was centralized was partial occupational history (any employment of several years or more) and cause of death (via, for example, social insurance records). With that much it might be possible at least to identify many of the industries which needed more detailed monitoring. Perhaps a program of licensing possibly hazardous work procedures could be instituted and the medical records of those working with them 'tagged.' What is absolutely *crucial* is a system of early warning and rapid response. The spread of a substance into thousands of uses is always accomplished in the place of other actual or potential ways of doing the same thing. Delaying this process (as well as its initial introduction) until safety can be demonstrated has to be less costly to both the economy and to human health than attempting to impose rigid safety standards after the substance has been thoroughly integrated into the industrial system.

Third, health studies must be conducted by researchers who are either impartial or, if anything, 'biased' towards worker health. That is, in practical terms they must either be part of a truly independent civil service or be in the direct employ of the unions. This is not to say that other research agencies should not exist as checks on the system, but rather that primacy in terms of staff quality, equipment, and ease of communication with decision-makers must rest with these bodies. I believe that the lesson of the history of asbestosis research, black-lung compensation and protection, and the vinyl chloride story is clear: research dominance should not rest with those interested in the relative slowness of changes in workplace practice. And new substances should not enter workplaces until they are proven safe.

Finally, there must be massive public education towards an understanding of delayed disease response and statistically patterned health impacts. No health or safety program can function without worker co-operation and even enthusiasm. This will not come about unless there is a generalized knowledge that one's life chances may well be affected even if one does not feel anything at the time of exposure to alleged dangers. That there are a few healthy retirees who did the same job for years proves nothing. Workers must learn to see the 'condition' of their industry (and their position within it) in terms of normal curves and probabilities.

NEW LEGISLATIVE EFFORTS

Ontario has recently passed a new occupational health act, Bill 139: The Employees' Health and Safety Act, 1976. In addition, the legislature now has before it another new bill, Bill 70, An Act Respecting the Occupational Health and Occupational Safety of Workers, 1978. Bill 139 allows workers the right to refuse to operate machines which they consider to be unsafe and allows the minister of labour to order the establishment of joint (worker-management) health and safety committees. Committees can be struck by the minister at the request of employees or if the minister feels the health and safety record of a given workplace warrants the creation of such a committee. Although the committee's powers are only advisory to the employer, this act, if pushed to its full potential by the ministry, might at least bring some issues to the attention of workers and management. It also grants employees the power to direct the Workmen's Compensation Board to deliver, via the employer, summary data on health and safety in the workplace.

Bill 70, which is expected to come up for second reading in the fall 1978 session of the legislature, goes a bit further in some of its provisions. The Resources Development Committee of the legislature amended the provision regarding health and safety committees to make them mandatory for all workplaces with more than twenty employees rather than leaving the establishment of these committees to ministerial discretion. This bill was further amended by the committee (which has an NDP-Liberal majority) to include *all* workers in the province under the jurisdiction of the bill. And further, this bill, in its original provisions (Section 21), requires that the ministry be notified of any new substance or combination of substances introduced into any workplace. The ministry may then require the employer to undertake an assessment of the effects of the substance. The ministry may also prohibit, limit, or establish conditions of use for toxic substances in the workplace (Section 20). This is a very large step to take and, if the act is passed and then stringently applied and enforced, Ontario's workplaces can be made much safer. Unfortunately the bill has a difficult time ahead in the legislature and particularly its toxics sections are open to a continuation of most of the abuses described in this paper. All is up to the ministry. But clearly Ontario may take steps that other provinces can then follow. In the process created by these two bills workers have structures available wherein they can learn to protect their health.

CONCLUSION

By way of summary and conclusion I would like to offer two brief sets of policy recommendations. It is hoped that the two items in the first set comprising the short-term minimal list would be politically possible, the

first on a national basis, the second perhaps at first within some relatively enlightened jurisdiction. The pending Ontario legislation does not
go far enough in this regard. In turn these changes could help to turn the
political balance towards further action. They are in that sense what
André Gorz would call 'radical reforms.'[48] The items in the second set, I
expect, will take considerably more time to bring into effect even with
the best of intentions on all sides. In most cases they will require some
greater level of political mobilization than will the items in the first set.

Some policy suggestions for the short term
1 / A revision of medical record-keeping techniques to devise ways to
allow for an on-going national correlation of illness and death records
with residential and occupational patterns. As was suggested above,
several co-operating nations could share out industries of particular
concern to each while collectively dealing with a relatively full range of
occupations. To keep initial costs manageable the process could start
with requirements that only specific diseases be reported (for example,
emphysema, silicosis, heart problems, cancers of all types). It is important to realize that in Canada, or any nation with a full health insurance
scheme, sufficient paper work is already done. The forms already used
must be modified, however, as must the procedures for recording information. Resistance will come from those within the medical system who
are not sympathetic to preventative medicine and from those corporations who may rightly fear a rise in workmen's compensation costs. I am
convinced, however, that the savings to society over the long term, in
health care costs alone, will make the initial investment seem very small
indeed.
2 / The passage of comprehensive 'right-to-know' legislation with regard to workplace health seems to me the *sine qua non* of upgrading
workers' understanding of the issues involved. It seems, regardless of
that goal, a minimum human right. Such legislation would include: (*a*)
The right to know the generic name of every substance present in the
place one works. This requires labelling *and* the distribution of lists to
each worker. Obviously the list could be the same for any given section
of a plant or a mine. The list must include both substances introduced
into the workplace and those generated by processes and equipment
used (for instance, dusts in mines). (*b*) The right to current data on what is
known about the substances or processes with which one works, written
in an intelligible way. Looseleaf guidebooks, regularly revised, could be
made available to the health and safety committees in each workplace.

The existence of such committees should be legislated. Much of this is already done in Saskatchewan. (c) The right to know about the present state of one's own health – or, in general, about the health of one's occupational group. This item perhaps belongs in a longer-term list given the habit of secrecy of some physicians. But health tests at the workplace will never be welcomed if the results are kept secret from the individuals tested.

Some policy suggestions for the longer term
1 / A general effort to lower TLV s where there is any evidence at all of health effects of short-term or long-term exposures. In many cases the levels should be set virtually at zero.
2 / The beginning of training programs in Canada for para-medical oc-cupational health technicians. An oversupply of nurses in some areas would make such a program possible on a crash basis. But in any case some medical school should begin such a program.
3 / Contract demands, especially in high-risk industries, for union-hired, company-paid occupational health technicians to do on-going health tests on all workers.
4 / The self-inspection of workplace conditions by the workers affected. Fixed metres and stations for monitoring hazard levels are not sufficient in most cases; too often they are deliberately placed in relatively low-reading areas. Portable metres are available for some hazards and those who are affected should be concerned enough to locate high readings when they exist. Obviously portable metres are not sufficient in them-selves either; nor can workers be trained to do readings and measure-ments for many dangers. However, this is possible in some cases, necessary in many, and in general could advance the level of concern. Both this item and the previous one are currently being undertaken by the auto workers (UAW) and the rubber workers (URW).
5 / Workmen's compensation payments should be constantly broad-ened (and are being broadened in many jurisdictions).
6 / The meaning of scientific 'benefit of the doubt,' as discussed above, should be revised as it applies to workplace (and environmental) health.
7 / Severely hazardous jobs – wherein worker death rates are clearly higher than those of the general population – should be either eliminated or qualitatively altered. Exposures can be eliminated by mechanical handling (grapple arms) or by means of separate air supplies. Other work tasks can be eliminated by automation (workers being retrained for other work with no loss in seniority). In addition, work with some substances

could be isolated from almost all other parts of a given factory and the hours and/or years of exposure of any given worker could be severely limited. In these cases such workers would have to be guaranteed continuing employment elsewhere in plant during the balance of their weekly work time and/or working lives. And many substances could be eliminated from present uses and substitutes found.

8 / All new substances should be thoroughly tested prior to introduction into non-laboratory workplaces.

NOTES

1 Stellman and Daum, *Work Is Dangerous to Your Health* (New York 1973), 172
2 *Ibid.*, 173–4; see also R. Doll, 'Mortality from Lung Cancer in Asbestos Workers,' *British Journal of Industrial Medicine*, XII (1955); I.J. Selikoff *et al.*, 'Relation between Exposure to Asbestos and Mesothelioma,' *New England Journal of Medicine*, 272 (1965); Selikoff *et al.*, 'Cancer Risks of Insulation Workers in the United States,' *Insulation Hygiene Progress Reports*, 6 (1974)
3 *Ibid.*, 174
4 Saskatchewan, Occupational Health and Safety Division, *Asbestos: Magic Mineral with Dust That Kills* (Regina 1974), 1; David Kotelchuck, 'Asbestos Research,' *Health/PAC Bulletin*, 61 (1974), 2–4
5 H.B. Holleb and A. August, 'Bronchiogenic Carcinoma in Association with Pulmonary Asbestosis,' *American Journal of Pathology*, XVIII (1942)
6 Oliver Bowles and F.M. Barsigan, 'Asbestos,' in *Minerals Yearbook 1951* (Washington 1954), 167
7 Selikoff *et al.*, 'Cancer Risks,' 1 (references referred to are itemized at page 4); Timothy C. May, 'Asbestos,' in *Minerals Yearbook 1964* (Washington 1967), 230
8 To confirm this one need only read the annual survey of new patents related to asbestos. These can be found appended to the articles on asbestos in the *Mineral Yearbooks*, for example pp. 232–5 for 1964. See also Selikoff *et al.*, 'Asbestos Air Pollution,' *Archives of Environmental Health*, 25 (1972), 1–13. The authors report a variety of studies of such diverse people as Finnish farmers, South African villagers, and urban residents of Britain and Pennsylvania. None of these persons had *occupational* exposures, but all turned out to have had exposures related to a variety of industries. A random 3000 autopsies of New York City residents indicated that 48.3 per cent had asbestos bodies present in their lungs. One need only realize that the brake linings of most automobiles now contain asbestos and these linings wear and thereby free asbestos into the air, that the heat and air conditioning ducts of many large office buildings are *lined* with asbestos, and that large sections of Lake Superior have been filled with it from iron-mine tailings deposited by the Reserve Mining Corporation.
9 Paul Biederman, ed., *Economic Almanac: 1967–1968 Business Factbook* (New York 1970)
10 Kotelchuck, 'Asbestos Research,' 4
11 *Globe and Mail*, Toronto, various dates throughout 1975. The *New York Times* of 19 Sept. 1974, pp. 1 and 27, reported the first century AD as the date of the earliest knowledge of the existence of an association between the two.

12 Environment Canada, *News Release*, numbered 136/12/15/75
13 The British have found that there are still health problems at this level of exposure.
    Selikoff has remarked on several occasions that in his view the only safe exposure level
    for asbestos is zero. This is not to say, of course, that the new legislation is not a step
    forward. For example, the *Globe and Mail*, 1 Jan. 1976, p. 8, reported that Quebec, the
    leading producer, '... has no official limits on the amount of asbestos fibre dust
    permitted in mines, mills, or factories.'
14 For a solid bibliography of Canadian labour history, see Gregory S. Kealey and Peter
    Warrian, eds., *Essays in Canadian Working Class History* (Toronto 1976), 185–94.
15 Many of the North American journals and magazines concerned with environmental
    issues written by scientists or scientific journalists for a more general audience were
    founded somewhat after this period.
16 A.J. Lanza *et al.*, 'Effects of Inhalation of Asbestos dust on the Lungs of Asbestos
    Workers,' *U.S. Public Health Reports*, 50 (1935), and discussion in Kotelchuck,
    'Asbestos Research,' 4
17 A.J. Vorwald *et al.*, 'Experimental Studies of Asbestosis,' *A.M.A. Archives of Indus-
    trial Hygiene and Occupational Medicine*, III (1951), 1, and discussion in Kotelchuck,
    ibid., 6
18 Two major studies which, among other errors, ignore the twenty-year latency period in
    lung cancer formation are discussed in Kotelchuck, *ibid.*, 6, 20–1, and 22–3.
19 Kotelchuck, *ibid.*, 23–4
20 P. Gross *et al.*, 'Pulmonary Response to Fibrous Dusts of Diverse Compositions,'
    American Industrial Hygiene Association Journal (1970), 125, cited in *ibid.*, 24
21 G.W. Gibbs, 'Some Problems Associated with the Storage of Asbestos in Polyethylene
    Bags,' *AIHA Journal*, 30 (1969), and *ibid.*, 24
22 Stellman and Daum, *Work Is Dangerous*, xiv, xv, and 4
23 Stellman and Daum, *ibid.*, xiii–xiv
24 *Ibid.*, 248. Much of the material in this section is based on Stellman and Daum and
    Franklin Wallick, *The American Worker: An Endangered Species* (New York 1972).
    But as well I have consulted here Anthony Tucker, *The Toxic Metals* (New York 1972);
    Lester V. Cralley *et al.*, eds., *Industrial Environmental Health* (New York 1972);
    George L. Waldbott, *Health Effects of Environmental Pollutants* (St Louis 1973); John
    S. Williams, *Environmental Pollution and Mental Health* (Washington 1973); and
    other sources cited below.
25 Stellman and Daum, *ibid.*, 256, 257; and from discussions conducted by the author with
    Paul Falkowski and other officials of the United Steelworkers of America and with Eli
    Martel (MPP, Sudbury) on various dates from from 1972 to 1974.
26 See particularly 'Special Issue on Inco and Sudbury,' *Alternatives: Perspectives on
    Society and Environment*, 2 (Spring 1973), and S. Berger, 'Pawns for INCO's Strategy,'
    Canadian Dimension, 9 (Oct. 1972), 24–34
27 Ross H. Hall, 'The Stack,' *Alternatives, ibid.*, 26
28 Regarding hazards to taxi drivers, see Ronald A. Buel, *Dead End* (Baltimore 1973),
    64–5; Stellman and Daum, *Work Is Dangerous*, 165
29 Wallick, *American Worker*, 4, estimates 150,000 coal miners *in the United States alone*
    have contracted black lung.
30 Stellman and Daum, *Work Is Dangerous*, 169. For Ontario, see the *Report of the Royal
    Commisssion on the Health and Safety of Workers in Mines* (Toronto 1976), particu-
    larly chap. 2

31 Stellman and Daum, *ibid.*, 182, 6–7
32 Quoted in Pat Gallagher and Don Kossick, 'Grain dust: Union Proves Health Hazard,' *Next Year Country*, 3 (Aug.–Sept. 1975), 10
33 Kotelchuck, 'Asbestos Research,' citing in turn *Washington Post*, 5 Sept. 1974, and *Wall Street Journal*, 2 Oct. 1974
34 B.F. Goodrich Company, *Press Release*, Cleveland: Chemical Division, 26 Aug. 1975
35 *Globe and Mail*, 27 Jan. 1976; see also public letter from E. Somers, director general, Environmental Health directorate, mimeo., dated 21 Jan. 1976
36 Wallick, *American Worker*, 2
37 *Globe and Mail*, 9 April 1975; Stellman and Daum, *Work Is Dangerous*, 130
38 Stellman and Daum, *ibid.*, 148
39 The figures in this section are all from Statistics Canada.
40 Department of Energy, Mines and Resources, *Mineral Policy Objectives for Canada* (Ottawa 1974)
41 Effects of these industries on the balance of payments can be quickly approximated by a look at C.J. Harris, ed., *Quick Canadian Facts*, 30th ed. (Toronto 1975), 98–100
42 See, for example, R.G. Smith *et al.*, 'Effects of Exposure to Mercury in the Manufacture of Chlorine,' *AIHA Journal*, 31 (Nov.–Dec. 1970), 687–700
43 See Kotelchuck, 'Asbestos Research'; in Canada, see, for example, industrial history sections of J.L. Granatstein and Paul Stevens, eds., *Canada since 1867: A Bibliographic Guide* (Toronto 1974), and also M.W. Bucovetsky, *A Study of the Role of the Resource Industries in the Canadian Economy* (Toronto 1973); John Porter, *The Vertical Mosaic* (Toronto 1965); and Wallace Clement, *The Canadian Corporate Elite* (Toronto 1975).
44 See, for example, E. Chinoy, 'The Tradition of Opportunity and the Aspirations of Automobile Workers,' *American Journal of Sociology* (March 1952) or the numerous more recent studies which have attempted to explain the so-called 'blue-collar blues.'
45 Some unions that have included occupational health issues as part of their collective bargaining efforts include the UAW (Autoworkers), URW (Rubberworkers), USA (Steelworkers), and the OCAW (Oil, Chemical, and Atomic Workers). These are large unions; many smaller unions have been unable for various reasons to proceed this far.
46 Two matters are worth reporting here: the OCAW for a time employed three full-time health professionals at their Denver, Colorado, headquarters; the UAW has made extensive efforts to train the workers themselves to monitor health conditions.
47 Wallick, *American Worker* – details of the study are given at p. 189; Leyton, *Dying Hard: The Ravages of Industrial Carnage* (Toronto 1975)
48 For a discussion of the term 'radical reform,' see Gorz, *Strategy for Labor* (Boston 1967)

7
The meaning of environmental problems
for public political institutions

C.A. Hooker and R. van Hulst

We take the environment to include not only the so-called natural environment but as well the socio-cultural environment in which we also live and move and have our being. (The 'natural' environment is itself rapidly becoming a human artefact and so falls more naturally into the second classification.) Environmental problems are usually thought to be problems of the natural environment: pollution, resource depletion, wilderness destruction, and so on. Problems in the socio-cultural environment are typically labelled 'social' and assumed to be different in nature and origin. In contrast, it is our view that, at bottom, the basic causes of our social problems are the same as those which generate problems in the natural environment. As we see it, our difficulties are rooted in a specific collection of values, institutions, and decision-making processes around which we have organized our culture.

We concentrate here on providing two case studies to illustrate our more abstract theses and at the same time make clear their practical implications. We have drawn these studies from the socio-cultural environment to emphasize our 'common root' thesis. They are the health care delivery system (HCDS) and Ontario Hydro. The first example is developed in more detail than the second, though neither of them can be but sketchily described in the space available.

In order to have some framework for understanding the case studies, we shall now offer a very brief outline of some of the motivating ideas behind our approach.[1] In particular we here focus on institutional structure and the decision-making processes which institutions embody. Institutions are created and peopled by persons, of course, and their values, beliefs, needs, and wants are the psychological foundation on which institutional function rests. But it is the design of institutions which is most directly amenable to political action. Moreover, the structural designs of institutions reflect the systems of goals, constraints, and incentives with which they operate. And finally, the values, beliefs, and wants people come to have are strongly influenced by their experience of the social institutions that surround them. So, without supposing that humanity can achieve utopia solely through institutional change, we concentrate on understanding the institutional origins of our environmental problems.

Roughly, Western industrialized society is organized around what we shall call the *market paradigm* (alternatively, the *neo-classical economic* paradigm[2]). Society is assumed to consist of individuals: autonomous, competitive (co-operative only when mutually advantageous), with a given set of preferences (wants). They are assumed to pursue the satisfaction of their wants through purchases of commodities in the commodities market, paying for these with money earned by selling their 'labour' to producers in the factors market. Markets and individuals are assumed not to influence each other except through buying and selling, neither do they hold any mutual responsibilities towards one another beyond proper conduct of buying and selling (but *caveat emptor*). This is the doctrine of 'consumer sovereignty.'

The implicit assumption behind this paradigm is that the substance of a high quality life can be acquired through purchases of commodities in the market. This assumption and the explicit assumptions briefly sketched above are perniciously false from our point of view. But instead of developing a detailed argument we shall simply assert two of their consequences. (1) The market paradigm fragments both the conception of the world and the bases for decisions in terms of market-priced commodities – setting aside the unpriced, unvalued rest – and market competitiveness ensures that consequences of actions not appearing in private ledgers are not taken into account. But reflection suggests that many of the characteristics of our environment which make for a high quality human life are collective, holistic systems features, which cannot be captured as collections of commodities. Hence, (2) the market paradigm would systematically generate socio-environmental problems through ignoring, or improperly conceiving of, these collective, holistic features (the 'externalities' of neo-classical economics). And market decision-making procedures, even when motivated by wider concerns, are still based upon a systematically misleading approach to problems. This is the basic skeleton of our abstract model and we believe that it captures the generation of both 'natural' and socio-cultural problems.

More specifically, we hold that the market paradigm leads to the development of a characteristic kind of institution with characteristic deficiencies which illustrate the systematically misleading approach to decision-making. Roughly, market institutions are designed to produce commodities. Such institutions we shall call *commodity intervention* (CI) institutions because their institutional designs are appropriate responses to the question: 'Given that the objective is to regulate the production of a given commodity, what is the most efficient design?' In CI institutions the natural notion of efficiency is that of efficiency of production, construed as narrowly as the production of the commodity permits. In consequence, CI institutions are usually organized as 'vertical' hierarchies. Information is maximally 'filtered' as it flows 'up' so as to remove all factors extraneous to the production-control parameters at each level; correspondingly, each decision is 'fragmented downwards' as it is specialized for the filtered factors at successively lower levels. Each level receives just enough instruction to perform its specialized production function and contributes only what is relevant to the production functions of others. The external relations among CI institutions model the assumed independence among commodities: each CI institution tends to operate independently of the others, relating only through market transactions involving the commodity it deposits into the market.

To support CI institutions is thus to make a variety of 'operating' assumptions about the preferred design of social institutions. For example, (1) satisfactory institutions are designed to produce commodities; (2) commodities can be defined, created, and used independently of the remainder of the social setting; (3) production of commodities can be exhaustively, objectively analysed and their design criteria determined entirely within the production process; (4) CI institutional roles are appropriate human roles. A little reflection upon the real societal impact of any commodity (motor vehicles, for example) and upon the human impact of work roles will surely convince the reader that these assumptions do not correspond to reality. None the less, they are a crucial part of the market paradigm – as integral as the sovereign consumer and the fully mobile nuclear family. The manner in which these suppositions conflict with human realities offers important insights into the origins of our socio-environmental difficulties.

The 'horizontal,' or commodity, fragmentation of CI institutions and their narrow commodity focus result in an incompetence to deal with the key holistic features of reality. However 'efficient' such institutions may be in themselves, what they do is often irrelevant, or worse. Moreover, the efficiency imperative leads to centralization (to concentrate control and to take advantage of 'economies of scale'), and to the attempt to transform all human issues into technical questions concerning production. In this fashion grow the massive, narrowly focused bureaucracies of 'experts' that remove public responsibility, initiative, and self-determination – in short, that undermine the political process.

As the last few phrases suggest, we hold these characteristics to be as true of public institutions as of private corporations. The former are often structured in the manner of market enterprises; institutional mandates are often solely conceived in terms of the creation and regulation of market commodities. Examples that spring to mind are: education, commodity = certified graduates; welfare, commodity = units of redistributed income; the health care delivery system, commodity = cures delivered; Ontario Hydro (or any other public utility in Canada), commodity = electricity. Even though our society does not confine itself to the market, our conception of society and our social practice have been dominated by the market ideal. Moreover, almost any public institution is, of course, heavily dependent on and consequently influenced by the private institutions that provide its technology, most of its management, and often its clients as well.

Given the pervasiveness of the CI model it is not surprising to find that our social institutions suffer from the same defects as private enterprises, in particular with respect to the quality of work roles and social relationships that they allow. In public institutions, too, we face the centralizing, technologizing, commodity-focused, public-excluding characteristics that market corporations display. As commodity 'solutions' to our problems become more and more entrenched and intractable, the public political process is increasingly undermined. It is these theses we hope to illustrate in the case studies that follow.

The very nature of our institutional problems makes it clear that further centralizing tendencies will only succeed in worsening the present situation. Meaningful public control, through the institution of iterative societal decision processes and effective means of conflict resolution, calls for decentralization and democratization. And there are institutional alternatives to CI institutions, institutions that are more process-oriented than result-oriented, that are explicitly structured in response to the question, 'What is the optimal design for maintaining the dynamic stability of a complex system?' One answer to this question is to choose the internal relations of the elements of the institutional structure so that they model, or match, those of the external system structure, for this is an economical way of supporting the flows of the relevant kinds of information and the formulation of the relevant kind of policy, that is, holistic, whole-systems policy. These institutions will typically not exhibit simple hierarchies because there will be complex feedback patterns which preclude such simple notions of authority. Put extremely briefly, we can characterize them as self-critical and self-designing (made possible by the feedback structure), exhibiting flexible 'cross-disciplinary' roles (because, for example, they deal with whole families and whole communities), employing 'multi-channel, multi-modal' communications and critical, iterative decision processes. It follows that these institutions are strongly decentralized, encouraging local diversity, yet integrated in the whole. The guiding model could be taken as a generalized model of 'plastic' biological control, for example, the neural coherency of the human body, which achieves a coherent (integrated) system while developing maximal local autonomy. Because they match themselves to the structure of socio-biological reality in order to aim at producing a match between that reality and a publicly chosen ideal pattern or design, we shall call them *match* (M) institutions.

Through a transformation to M institutions we would hope to travel some distance towards restoring quality solutions to socio-environmental problems, quality social roles for people, and a revitalization of the public political process.[3]

HEALTH CARE DELIVERY

The organizational structure of the HCDS in Canada is dominated by two institutions: hospitals and private offices. Financially the system can be characterized as a subsidized entrepreneurial model: hospitals and health personnel are subsidized by government, but operate on a fee-for-service basis. While a government-subsidized comprehensive medical and hospital insurance plan covers people in case of illness, left to private enterprise are medical insurance, hospital construction and the provision of drugs, medical supplies and instruments. It is generally agreed that we enjoy a high standard of medical care.[4]

Yet there is no reason for complacency: the ills of the present system are widely perceived, not in the least by many among the medical profession itself.[5] A study of the ways in which a system fails is both informative to one who wants to understand the system, and relevant to one who wants to improve it. We shall therefore proceed to discuss some common problems and concerns, try to analyse these within the framework constructed above, and then suggest some ways of overcoming them.

A first source of concern, be it mostly to government officials, is the rapidly rising cost of medical care without a substantial concurrent improvement of health.[6] The annual cost escalation has been between 12 and 16 per cent, far in excess of economic growth. Furthermore, government bodies have not been very successful in inducing health workers (especially general practitioners and specialists) to establish in smaller population centres and in the more remote parts of the country. As a consequence, there is a maldistribution of health care over the different parts of the country. There also are too many acute-care hospital beds in high technology hospitals, especially in the larger cities, but too few chronic-care beds and too few beds in low technology community hospitals. Costs of drugs, dental care, and other items not in the insurance plan are often unaffordably high. Complaints about the doctor-patient relationship are common: patients resent the impersonal treatment they often receive; they feel that they are being dealt with as cases rather than as persons; they feel frustrated over the paternalistic and moralistic behaviour that some doctors exhibit and about the general tendency to professional 'mystification.'[7]

For many health workers the rigidly hierarchic structure of the HCDS and the virtual lack of feedback and assessment of quality of care are sources of frustration. The fragmentation within the HCDS is apparent to both professionals and the public: there is much unnecessary repetition of diagnostic tests, a lack of integration between HCDS s and other social services, and, most importantly, it is extremely difficult for the medical profession to establish an effective program of preventive medicine, which necessarily relies on many other institutions of society.[8]

Some concerns are less widespread than the ones mentioned above but are possibly of an even more serious nature. Evidently, the number of unnecessary operations is a function of the number of surgeons and acute-care hospital beds available.[9] Most accidents involving children do not happen in the streets but in hospitals.[10] In general, iatrogenic (that is, doctor-made) diseases are frighteningly common and many patients receive wrong or unnecessary drugs.[11] The effectiveness of many standard medical techniques is doubtful or has never been clearly established.[12] Patients are often denied a death with dignity surrounded by those who are close to them and are on occasion made to suffer severe anguish and pain through prolonged and futile treatment.[13]

In short, all is not well with our medicine, in spite of good intentions and the many wise and devoted people in the health care professions. The suggestion we wish to explore is that many of these difficulties may be illuminatingly presented when traced back to the institutional structures that underlie them and more clearly understood when these structures are in turn traced to the market/individualist approach to medicine.

How, then, are our health institutions structured? Are hospitals not predominantly organized to provide cures with the greatest technical efficiency? Are pharmaceutical companies not structured to sell drugs? Is not the role of the general practitioner becoming increasingly confined to providing symptomatic relief and referrals to technical specialists? Is not the food industry designed to market maximally profitable edibles? To these activities we might contrast, respectively, preventive medicine and personal care, lifestyle therapy, family and community therapy, and nutrition designed to promote health and enjoyment of food.

We are not suggesting the elimination of technical expertise and technology, which we agree are necessary and important. What we are concerned with are the less tangible, less easily priced qualities ingredient in a rich and wise approach to health which tend to slip unnoticed and unvalued between the specialized functions. Our concern is with the structural inability of these institutions to concern themselves with anything but efficiency and growth in the production of some commodity which they market – drugs, cures, TV dinners, and so on. An institution that 'cares' for sick people by only attending to their physical illnesses has to ignore the fact that the majority of diseases have a socio-environmental origin. It has to ignore the fact that diseases are not merely technically definable malfunctions, and that, for example, nutrition and freedom from emotional stress are intimately related to biological health. In fact, the whole pattern of society strongly influences the pattern of ill-health.

The general problem, then, is that the HCDS functions as if it were fragmented into institutions for the production and delivery of individually marketable commodities, whereas health is a holistic feature of a total pattern of life in a given environment.

A nice illustration of the problems of 'commodity-fragmentation' is afforded by the 'paradox of the motor vehicle.' In our society both the provision of motor vehicles and of hospital emergency services are regarded as commodities central to a high standard of living. In consequence, society has invested much of its substance to make both increasingly available. But since the motor-vehicular transportation system fragments moment-to-moment control among its individual drivers, it leads systematically to accidents. Thus the *collective* result of the supply of these two *individual* commodities is to put an increasing number of people through the drive-accident-emergency-recovery-drive cycle and to do it increasingly efficiently.

We shall now briefly discuss some of the ways in which the social and natural environments influence man's health. It has been established that over 90 per cent of all deaths in the United States have significant environmental contributing causes.[14] Although comparatively little research is devoted to preventive medicine, there is clearly wide scope for its application. In fact, many illnesses for which we do not now have a cure can nonetheless be prevented.[15]

Some threats to human health are at least partly due to a self-imposed lifestyle: drugs, alcohol, tobacco, stressful occupations, poor nutrition, lack of physical exercise, and so on. Two comments need to be made. First, the choice of a lifestyle is not usually a conscious one, nor is it made in a social vacuum; to the extent that social pressures promote the adoption of an unhealthy lifestyle, effective prevention will require social changes. Second, our society has a long tradition of placing all the blame on to the individual for what in effect amounts to accommodating himself to social circumstance – a tradition supported by the myth of 'consumer sovereignty.'[16] Surely the great *political* task has always been to aid in the development of an ever more humane culture and this mandate cannot responsibly be avoided by false appeals to consumer sovereignty.

A second category of socio-environmental factors are those affecting health but over which there is little or no individual control. Under this category fall such factors as pollution, destruction of recreational land, declining food quality, the hazards of food additives, medical care and transportation, limitations of education, and so on. The health care professions have typically avoided confronting these issues, choosing instead to regard them as the responsibility of some (other) government agencies. This attitude is consistent with their restriction to the commodity: cures. The consequence is the proliferation of 'external diseconomies' typified by the paradox of the motor vehicle. The institutional organization of health care in our society admittedly makes it very difficult for health workers to intervene effectively in the processes which create such adverse socio-environmental factors. Neither their training nor their equipment nor accepted societal roles (all of which are strongly influenced by corporate policies[17]) is conducive to critical action in these realms. And, furthermore, when such action is attempted it is bound to lead to serious conflicts, because optimal community health and optimal corporate profits tend to be irreconcilable goals. The resolution of such conflict is indeed a matter of public political action. To delineate the hazards and to draw attention to environmentally induced health problems, however, is clearly the responsibility of the medical professions.

All of this has left us in the paradoxical situation where the potentially most effective field in medicine – preventive medicine – receives relatively little attention. The role of preventive medicine seems currently limited to technically trivial treatment (for example, routine immunization, however important), preaching, and delegation of responsibility to other groups or agencies. Moreover, much of what goes under the name of prevention is focused still on the individual and is more properly described as early detection and control rather than prevention proper.[18] The bulk of medical resources are at the present time focused on early detection programs, for example, for cancer, and relatively few resources are devoted to the study of the larger socio-environmental factors implicated in the illnesses, for example, nutritional standards required to avoid cancer of the colon.

We have tried to illustrate one of the consequences of a cI organization to health care institutions by pointing out how the commodity approach clashes with the long-term maintenance (preventive) approach. The one *reacts* to *individually defined defects* with the provision of individual *commodities*, the other *pre-acts* on the *collective social condition* to achieve a *collectively defined* pattern of *health*.

Another consequence of cI institutional design is the emphasis upon technological control to achieve production efficiency. Its most important appearance in the health field is in the transformation of health concepts, problems, and procedures into technologically defined concepts, problems, and procedures. A related consequence is that society becomes divided into the technical 'experts' making purely technical decisions (that is, doctors and the like) and passive consumers (patients). The market model is that of the consumer, who has no responsibility to define the production of commodities, but is assumed sovereign in the choice among them. Yet surely it is clear that this model does not, and ought not, apply to health care delivery in a humane society.

The market approach would perhaps be tolerable if the delivery of health care were a purely technical activity such as the supply of water or the production of motor vehicles (though we should argue its defects even here, in similar terms to the social impact of the motor vehicle). But surely health care delivery is not. A condition only becomes a disease after a society has labelled it as deviant; all talk of illness expresses social value judgments.[19] Similarly for health. Comparative social studies confirm this. So does, for example, a glance at the four definitions of health appearing in one well-known text.[20] Health is defined as: (1) the lack of positive impediments to a person's functioning or survival, (2) soundness of body: that condition in which its functions are duly and efficiently discharged, (3) a state of complete physical and mental well-being, (4) conformance, within normal limits of variation, to accepted standards of health criteria for age, sex, community, and geographic location. This last definition, which is perhaps the one most widely used, clearly illustrates the tendency to substitute a technical definition (the medical model of the body social) for an evaluative one and, in the process, the tacit appropriation by the experts of the normative component (the setting of standards by the medical profession). Moreover, this is often accompanied by the suggestion that these judgments are objective, not normative, because scientific.[21] In this fashion the patient is denied responsible participation in his own health decisions, an experience which carries across to the larger health-determining pattern of his life.

Another characteristic of CI institutions is to respond to 'inefficiencies of production' by centralizing the production processes and their control. For example, in a technological model of health care as commodity (cure) production, medical information becomes purely technical information and, as such, an obvious target for centralized 'consolidation,' especially when the current institutional (commodity) fragmentation within the HCDS leads to information-processing inefficiencies. Hospitals, for example, often have to repeat diagnostic tests on a patient that were already performed elsewhere. It is the physical storing of medical records in a form that makes them accessible to all of the component institutions in the HCDS that creates problems. The solution most favoured by medical technocrats is storage in a centralized computer system to which all health care personnel have access. But this would give this professional class enormous potential power. Recall that patients are already often denied their own medical information on a doctor's judgment of what is in their own interest, and that an increasing number of social privileges – from executive jobs to credit extensions – are influenced by medical records. The centralized solution amounts to a 'Faustian bargain' in which the public surrenders control over the conduct of life in exchange for expert care. This calls to mind the similar proposal in the centralized nuclear energy field. In that field we already face serious proposals to curtail public discussion on certain crucial matters because it is not 'in the public interest.' Compare also the recent Ontario experience of the denial of information to the public in the Port Hope radium contamination problem.[22] The time may not be far off when medical information is denied the public in the same manner. In this fashion CI institutions make a self-fulfilling prophecy of the claim that only tight centralized control is capable of coping with our problems.

There are some obvious alternatives to the centralizing response. Individuals may be made responsible for keeping their own medical records, or medical records may be kept in local community centres, made easily accessible to each individual but accessible to the profession only with the permission of the person concerned. That these kinds of ideas are hardly considered, though they offer the same kind of integration of information for which the centralized solution is sought, shows how strongly our imaginative social horizons have been confined by the CI model.[23]

Finally, we consider briefly the role and consequences of competition in CI institutions as it applies within the HCDS. Theoretically, competition is what ensures production efficiency and, allied with consumer sovereignty, that consumer demand will be met. In a world dominated by the private ownership of technical commodity-production processes, it reinforces the CI institutional focus on commodity production to the exclusion of wider responsibilities. This is manifest within CI institutions by the appearance of strong competitive hierarchies organized about the control of production. The HCDS is no exception.

First, consider again the over-all institutional design of the HCDS. As we remarked, it is dominated by large centralized hospitals and private practitioner's offices. This structuring can be understood as a natural outcome of its market orientation, for the entrepreneurs of the HCDS are the doctors and they fundamentally sell advice-directives (corresponding to the private office) and technological curative processes (corresponding to the hospital). That hospitals tend to be very large and centralized while doctors' offices are small and essentially unequipped can be understood as a natural outcome of the achievement of maximal efficiency in the market by exploiting economy of scale. The structure may be compared in this respect with that of almost any other major business enterprise. (There is a third component to the system, the private medical laboratory, but this only serves to complete the parallel between the HCDS and the typical structure of private industry.)

We also find competition among health personnel throughout the system. This competition can take many forms, typically manœuvring for wealth (income), power/control, security, prestige, and so on. (We may include in the competitive arena private medical laboratories, private health schemes, and the pharmaceutical industry.) It is characteristic of CI institutions that vertical position in their hierarchies carries with it wealth, power, and security simultaneously. Any individual faced with a decision involving larger humane values and responsibilities, but operating in such a context, must face overwhelming pressure to acquiesce to the narrow production outlook of the institution in which he must survive. The division of hospital beds, for example, between primary care beds (for minor cases), secondary care beds (for secondary referral) and tertiary care beds (for difficult cases) is usually achieved by competition between general practitioners, specialists, and academic super-specialists. It is clear, however, that these decisions are as apt to reflect the competitive aims of the personnel involved as they are to reflect the public interest. Again, the technical 'programming' of human births in large hospitals, through anaesthesia, induction of labour, etc., largely in the technical interests of its expert staff and for the administrative convenience of a large ward, illustrates the substituiton of institutional interests for wider community values.

The joint influences of competition, technological emphasis, and centralizing response to inefficiency are nicely illustrated by the recent controversy surrounding the closing of regional hospitals in Ontario in order to remove excess hospital beds.[24] There is little doubt that competition among hospital administrators, specialists, and municipalities for scarce resources (for example, finance, patients) leads each local group to seek to establish the largest and most technologically sophisticated possible medical facilities in advance of other competing groups. Secondly, competition among individual health personnel evidently leads to the same result, for there is substantial evidence to the effect that the number of medically discretionary operations rises in direct proportion to the available hospital beds.[25] On the other hand, it is clear that such a situation will not in general benefit the public, both because its members will be subjected to an increasing number of largely unnecessary operations and because the cost of the medical system increases very rapidly as the number of hospital beds increases. From the strict CI point of view of technological/economic efficiency the relevant response is clear: close down the smaller 'outlets' in favour of larger, centralized 'plants' and so capitalize on the economies of scale. This is what seems to have happened in Ontario, and from a managerial point of view the move was probably justified. But in the process local communities lost their local services, the prospect of achieving a community-involved, person-responsive, services-integrated HCDS (see below) was dealt another blow, the trend towards undermining the local rural community and accelerating the growth of megalopolis reinforced. To understand the values and lifestyles over-ridden in such decisions, and to be able to perceive the possibility of a feasible alternative, we need a different institutional paradigm from that of the CI institution.

Finally, the opposition which is clearly currently felt in the Canadian HCDS to any policy which involves shifting its institutional structure strongly towards community-centred, preventive-oriented, low-technology institutions (for example, through community medical centres and similar intermediate institutions) clearly arises from the competitive market structure of the HCDS: the move is resisted by hospital administrators and specialists because they see such intermediate institutions as being in competition with the large hospitals and the move is equally resisted by the private practitioner primarily because he sees it as a threat to his entrepreneurial independence (whether directly through the participation of the community in the operation of its local health programs or indirectly through his loss of privileges in the large centralized hospitals which function as centres of information and power in the current medical establishment).[26]

Other examples of the maldistribution of power in medical institutions abound. Hospitals are well known to be rather strictly organized on hierarchic principles. The real power is virtually concentrated with the surgeons who decide about patient intake, length of stay, and allocation of beds. Other health personnel (administrators, nurses, etc.) form a descending linear hierarchy below them, each maximizing its advantage in the institution according to position in the hierarchy. It is equally well recognized, however, that the interests of these various groups, and that of the patient, are often opposed (for example, the humanization of nursing roles is in conflict with the interest in efficiency held by both specialists and administrators). Who should decide in these conflicts of interest and in what societal framework should the decisions be made?

Also, through their specialist knowledge and high earnings the pharmaceutical companies wield an enormous amount of power over doctor and (hence) patient. Since their products serve a roughly constant demand – indeed, a diminishing demand to the extent that preventive medicine is sucessful – but their profitability as CI market institutions depends upon ever increasing sales volumes, their interests are not always consistent with those of either doctors or patients.[27]

In this connection, it is a more subtle reflection of the general market form of our culture that we do not possess the kind of public institutions which would make for a public, preventively oriented, morally responsible discussion of medical information and practice generally and the supply and use of drugs in particular. It is characteristic that in our society, besides inaccessible academic journals and the occasional pamphlet in the private practitioner's office, the only publicly accessible supply of information in any way related to public health is that to be had through the advertising media. The desire of advertisers to change the public's behaviour in their own interests has often resulted in the substitution of an irrational and fantasy-ridden pseudo-science for the real thing. In our culture people have been persuaded that, ignoring the larger pattern of their lifestyles and living conditions, it will enhance their health if only they will surround themselves with a wide variety of medical commodities, from vitamin pills to incidental cosmetics (vaginal deodorants and mouth washes) many of which have been shown to be medically useless or worse. They have been persuaded that a visit to the doctor is unsuccessful unless he has been persuaded to prescribe a drug of some sort or that their daily stress or particular problems are most appropriately treated by some non-prescribed drug that temporarily removes symptoms and so on. The upshot of this miseducation is that people fail to understand the real determinants of their own health, misidentify the causes and nature of their medical problems, and, more subtly but more importantly, come to identify healthiness with the provision of a set of visible commodities.

Nowhere are the results of this miseducation clearer than in the field of food and nutrition. The natural ability of man to determine by appearance, taste, and smell whether a food is healthy and nutritious has been removed by the food industry with its profuse use of artificial additives of all forms (colours, tastes, smells, etc.) and of refined carbohydrates.[28] The concept of healthy nutrition has been undermined in the public mind through lack of information and it has been confused by a heavy advertising overlay on convenience, novelty, and appearance. Further confusion results from such advertising practices as singling out refined carbohydrates as good sources of energy while omitting reference to the nutritional context in which the healthy provision of bodily energy occurs, or by the suggestion that basic items such as bread are successfully and better enriched through the artificial injection of a few minerals and vitamins – after the original product has been thoroughly depleted through processing. This too is a natural outcome of the generalized market structure of our culture. We can hardly expect a sugar company to point out that a diet high in refined carbohydrates can lead to dental cavities, heart disease, and diabetes. As remarked, it is characteristic of our society that these forms of information and misinformation dominate the publicly accessible forms of health education.

The education of health workers reflects in many ways the convictions and structural characteristics discussed above: vertical hierarchy and fragmentation, competition, orientation towards cure rather than care, divorce of ethics from action, stratification of power, etc. For systematic market reasons, bolstered by a variety of historical circumstances, medical education is centred upon hospitals, and the curriculum of medical schools has been traditionally organized with the specialist primarily in mind.[29] Until quite recently comparatively little attention was paid to preventive medicine. While the present trend away from this circumstance is encouraging, it should be noted that the higher income, power, and status of specialists, and the socially central role of hospitals in the HCDS, continues to place severe constraints on the serious modification of education curricula. Moreover, it is still the case that most medical students are drawn from a tiny group of biochemists and biochemically oriented medical specialists. The effects on human health of such environmental factors as nutrition, pollution, social stress, and so on still receive relatively little attention in medical education. Political, ethical, and societal aspects of medical organization and practice tend to be underemphasized in medical schools (both within the curriculum and outside of it). Medical schools have only recently made any serious positive effort of their own to recruit quality medical students whose interests and background differ appreciably from those aiming at conventional entrepreneurial medical roles. (And only a few Canadian schools have taken this initiative at the present time, at that.) Finally, continuing education for doctors is dominated by information provided by pharmaceutical companies and the like which, by its very nature, is narrow in outlook. And while the education of nurses, occupational therapists, and other allied HCDS personnel has often been broader, they are not influential enough in the HCDS hierarchy to escape the narrow institutional roles it assigns them (see our discussion of alternative roles below).

The structure of the Canadian HCDS and the social role of the medical professional have recently been the subject of much discussion and a number of government reports have appeared with explicit recommendations for change.[30] The general gist of much of the discussion and of the reports is that the time has passed when the medical profession can make its decisions solely on the grounds of efficiency rather than efficacy, that providing cures alone does not bring one closer to a healthy society, and that the medical institutions must become more integrated both among themselves and with the rest of society.

Significant and commendable as these conclusions are, the recommendations for practical change issuing from these reports are, it seems to us, still not sufficiently penetrating and thorough. For if, as we have tried to demonstrate above, the root of the present problems in the Canadian HCDS is a fundamental inability of the institutional structure of the system to allow it to follow any other than a commodity-oriented approach, it follows that nothing but a penetrating and thorough transformation of the institutional structure will be effective in providing high quality health care.

What do the practical recommendations of the reports look like? The Lalonde report recommends separating primary and secondary care, with emphasis on the former and particular care taken to achieve effective co-ordination and integration between the two. In order to facilitate planning and operation at the local level the report proposes instituting a local District Health Council and an Area Health Services Management Board, together with some other organizational changes at the regional and ministerial level. These recommendations can be read as an important move in the direction of decentralization and community orientation. Unfortunately, it is possible, and likely more accurate, to read them as fitting a more thoroughly managerial model. The primary health care sector becomes, on this view, a mere referral service and in fact a service likely to be more efficient in recruiting patients for the secondary or tertiary treatment level than is the present operation of the HCDS. Under these circumstances, the basic character of the HCDS would be unchanged and, moreover, it would be difficult to see how the primary care level could continue to be the emphasized sub-system as the report evidently desires. (We note that the Ontario minister of health has already rejected these attempts to decentralize the planning and operational structure of the HCDS, apparently his only explanation being that hospitals fear the institution of 'yet another local bureaucracy.')

This same schizoid combination of community-oriented intentions cast in a managerial/medicalizing approach to health care applies equally to the conclusions of the 1974 Science Council of Canada report, which recommends that health care in Canada be reorganized into an integrated system that is decentralized (organized around community clinics rather than large hospitals) but interconnected through a computer network linking together records for all birth, ambulatory care, hospitalization, and death in Canada. Though there are other components of the report we would single out for praise as well as blame in a more lengthy critique, our main criticism of the report has to do with its tendency to suppose that the improvement of quality in health care centres about provision of such managerial tools as the computer rather than about the redesign of the societal relationship between medical professionals and the community, especially the local community.

In the limited space remaining to us we can hardly present a detailed discussion of an alternative Canadian HCDS. Instead, we shall largely confine ourselves to brief general comments on the characteristics of the kind of alternative HCDS for which we have been arguing.

The first criterion by which to judge any proposed structural change in the HCDS (as in any other institution) is whether it does or does not enhance the institution's own potential for self-conscious design in sensitive interaction with the systemic structure of the environment. Is the institution responsive to the pattern of life in the community? Can it adapt its procedures to the changing needs of families and individuals within the community? These characteristics hold only in a very narrow technical compass of our present HCDS and are not true at all of the theoretical CI institution (since it entirely lacks relevant feedback).

Secondly, it is our opinion that, rather than the further transfer of responsibility and power to the medical profession which the reports referred to above tacitly advocate, only a deep understanding by the community of the social, political, moral, environmental, and biological aspects of health and disease can provide a proper basis for optimal human function. To generate such an understanding requires much more than instituting community clinics and preventive health programs as these have been traditionally understood; it requires a determined effort of everybody in the health institutions, the government, the educational system, and the public media to make the concern for optimal health and well-being the responsibility of the community rather than that of the doctor or of the individual. The initial stage of this transformation resolves itself into a series of political issues concerned with changing the structure, responsibilities, and (so) functioning incentives of the institutions involved. For example, government might create alternative medical education institutions (or create the incentive to create them), linking them more directly with other social planning and service agencies. Or government might choose to support more vigorous local media, encouraging them to become involved in local medical education, in local community health issues, and in laying the foundation for a serious community program of preventive medicine.

It is quite generally agreed that neither the structural model of extreme diversification, in which there are only isolated doctors and other personnel, nor the model of exteme centralization, in which all services are concentrated in a few gigantic institutions, is socially acceptable. We have suggested that no mixture of these represents a significantly more acceptable solution than does either of the extremes. Instead, we believe that the Canadian HCDS must develop a range of new institutions, small-scale and integrated in an effective way with other local, community institutions in order to deal effectively with the entire local community. We want to aim at a community-oriented institution exhibiting a biological model of 'plastic' control, that is, one in which there is maximal local autonomy combined with selective integration. In addition, our remarks imply that the HCDS must develop important functional roles in other societal institutions, for example, transportation, education, energy, and environment.

A simplified sketch of the final distribution of institutional components in the HCDS might run somewhat as follows: the major emphasis of the health care *delivery* concentrated in integrated, community service institutions, backed by specialist secondary and tertiary care institutions (on a scale appropriate to real need in a healthy community, as defined by the community), the major emphasis in health *design* to appear in inter-institutional relationships between the HCDS and other major societal institutions, the whole backed by the minimal administrative bureaucracy necessary for its effective co-ordination.

Out of all these components, we choose only to comment briefly upon the integrated, community institutions. First, we emphasize the integration of community functions in one institution. In our conception, the community centre would include a library (through which a wide range of medical information could be dispensed), baby health clinic, nursery and day care centre (which together provide for continuity of care for the growing child and continuity of education in health care for the parents), a family doctor centre, (an equivalent of) extended out-patient services (perhaps with limited facilities available for short stays, for example, for a family living in with a sick young child), home medical services (nurses, paramedical personnel with portable equipment, visiting doctors, etc.), dental service (with emphasis on preventive dental care), recreation facilities and environmental quality control (again concerned with promoting general community health), employment and welfare services, and a local education board (for adult education as well as that of the child, with strong citizen participation in each). So constituted, such centres can be activity and learning centres for whole families, thus reducing the sense of monolithic dominance and sense of patient helplessness that modern hospitals tend to create in the public. They can permit flexible family arrangements in the care of the sick, the pregnant, and the like, thus reducing the emotional trauma suffered especially by children under the existing system, prepare adequately for the process of formation of a family, encourage self-care through information and familiarity with medical procedure, and so on.

Such small-scale institutions promise the breaking down of professional barriers and the merging of roles so as to permit a coherent treatment of people and the community along the systems-sensitive lines that really matter; they therefore will contain persons whose present skills would be designated as those of doctors, nurses, social workers, occupational therapists, family counsellors, urban planners, etc., working as integrated teams.

It is inevitable, then, that these changes require a different, and broader, education for all of these onetime specialists. This is equally true of those medical professionals who must work with regional industry, urban planning, energy and environmental planning, and play other such inter-institutional roles. Just as we have advocated the transformation of the institutional structure of the HCDS, so there must be a corresponding transformation in the structure of those educational institutions that educate (currently 'train') personnel for these institutions.

In short, it is not without a major institutional reformation that our aims can be met. Of course, it is our hope that such a reformation can be accomplished through a series of incremental steps with particular care being taken not to estrange or needlessly damage any group of society, inside or outside the current HCDS. Moreover, each step will require careful consideration of the real constraints which preservation of high quality, technical medical care – and economic reasonableness – place on institutional flexibility (though these constraints should not be over-inflated, as they tend to be at present). It is significant, though, that at the present time there seem to be very few workers actively concerned with planning effective incremental change along the systems-sensitive lines that we have advocated. In the long run, our success in these matters will depend very much on the extent to which we are able, as a society, to replace individual goals by communal ones compatible with the preservation of human values.

These goals define the framework for a program of political action. The general implications should have been clear enough as we proceeded through the analysis. To spell out a practical, step-by-step program, say for Ontario, would be contrary to the spirit of our argument: if the provision of health care is to be a communal responsibility, rather than the exclusive concern of a small group of 'specialists,' then the development of new institutional forms for the HCDS calls for an iterative process of public discussion, action, and learning. We can only hope that these remarks will help to stir a serious debate.

And we can present other illustrations of our approach in the hope of crystallizing the issues. The issues are even sharper in the energy field, to which we now turn.

ONTARIO HYDRO

Ontario Hydro is the second largest publicly owned utility in North America with roughly 22,000 employees and assets worth over $5.5 billion. It supplies electric power to the province of Ontario which has one of the highest per capita electric energy consumption rates in the world: approximately 9000 kwh per person per year.

Ontario Hydro has received wide authority in relation to the generation, transmission, and distribution of electric power (and indeed of all forms of municipally managed energy) throughout the province. It also has regulatory and quasi-judicial authority over other electrical energy producers and local electrical utilities. Rates are set by Hydro, subject to review by the Ontario Energy Board.

Ontario Hydro's mandate has traditionally been 'to provide power at cost.' It would be characteristic of a commodity intervention (CI) institution to interpret this mandate narrowly in a market-production sense. Hydro's understanding of its mandate is 'the provision of generation and transmission facilities to supply the electrical power demands of the people of the province.'[31]

But in fact one cannot reasonably interpret the mandate in this narrow commodity-oriented manner. For example, even the 1972 report of Task Force Hydro specifies the objectives of the corporation as follows: to meet demand at lowest feasible cost; reliability; to support regional development policies; to stabilize business cycles; to support environmental policies; to exploit new technologies; and to stabilize capital markets. An encompassing list of goals such as this is already difficult to reconcile with the frequent admonitions, both from government and Ontario Hydro officials, to have government intervention in the corporation's market activities held to a minimum. Moreover, and more importantly, one wonders whether it can be assumed, as Ontario Hydro obviously does, that the demand for electric energy is independent of government and utility policies. This assumption is characteristic of a CI institution. But if it is false, then surely the determination of future generating capacity should be a matter of general public policy, rather than a technical judgment of a power-producing utility company.

Before examining these questions a quick historical survey and a glance at future options are in order. Until fairly recently (1960) Ontario Hydro relied on water power for over 99 per cent of its total power generation, but in the 1970s this has dropped below 60 per cent. The implicit policy of maximizing sales of electricity (by regressive rate structures, advertising, etc.) led initially to important economies of scale: tapping hydro power involves only capital investments for the building of new dams, relatively little maintenance, and no fuel costs. Moreover, sites that were developed before 1960 were all close to the important load centres. When by 1960 the most attractive hydro power sites had been developed, and with the market for electric power still growing at around 7 per cent per annum (corresponding to a doubling time of roughly ten years), new generating capacity had to be developed in the form of thermal power stations. Initially these used fossil fuel but more recently nuclear power stations have been added and future proposals call for the latter type to dominate further development.[32] These power stations consume large amounts of fuel, the price of which has been growing rapidly. Nuclear power plants, even more than their fossil-fuel counterparts, are highly capital-intensive, require a large operating staff, and have high maintenance costs. Moreover, both types of plants are sources of various pollutants (sulphur dioxide, ash, radioactive elements, heat) and both constitute health and safety hazards for an increasingly large proportion of the population.

Under these circumstances it is understandable that both Ontario Hydro's pricing policies and its high load forecasts have been increasingly seriously questioned.[33] Critics have pointed out that much energy is wasted and that an intelligent conservation program can cut down electrical power consumption considerably. They have also noted that the provision of increasing amounts of power has serious societal impacts and they have asked whether the time has not come to reconsider this aspect of the 'growth ethic.' Hydro's position in response is simple: its mandate is to meet demand, the demand itself is simply a fact (an action, clearly, by 'sovereign consumers'). Future load growth (at the historical rate of 7 per cent per annum) is repeatedly described as 'inexorable' (without proper explanation) and scenarios for the future reveal an alarmingly inflexible commitment to the original vision of a vast network of increasingly large nuclear generating stations.[34]

Ontario Hydro is institutionally structured around the construction and maintenance of its generation and transmission technology. Policy is determined for the entire organisation, and hence for the province, by a small executive group at the head of the institution. Below this level, there is minimal discussion concerning *policy* issues between Hydro and other segments of society, in particular with government agencies and municipalities (though there is undoubtedly much discussion of technical detail). Similarly, public input to policy decisions has until very recently been essentially nil. Recently, public hearings have been established for decisions concerning plant siting or transmission line corridor siting. However, the hearings have often been held at such a late stage of the planning process that public input can influence only relatively minor details of the proposal. Public participation in the over-all policy-making process for the future development of energy systems in the province is still almost non-existent.

In short, a familiar picture emerges: that of the centralized, technically oriented institution, efficient in the production of its commodity, but structurally unable to relate to the wider social ramifications of its activity.

One of the first things that strikes one when considering Hydro's mandate is that it is limited to the provision of just one form of energy – electrical energy. Clearly this form of energy is being treated as a separate marketable commodity. But there are many complex and socially important trade-offs to be made between electrical and other forms of energy, in respect of security, safety, thermodynamic efficiency, pollution, and so on.[35]

Electrical energy is a particularly high-grade form of energy and it can therefore be used in nearly all applications. Heat, especially low temperature heat, on the other hand, has only limited applications. We live in a world where low-grade energy abounds but where high-grade energy can only be generated at fairly low efficiencies. The majority of the energy we consume, however, is used for low-temperature heating and cooling. It seems obvious, then, to utilize the abundant, and renewable, low-grade energy for these uses; hence the current interest in solar heating and cooling, heat pumps, and like devices.[36]

The increasing concentration in Ontario on high-grade generation technology, especially of electrical energy, stands in striking contrast to this approach. There is practically no support for the development of low-grade energy technology. It is characteristic of Hydro that, in its 1974 paper, *Long-Range Planning*, solar energy is mentioned only in passing and then only to refer to *electric* solar cells (which are indeed still expensive and poorly developed). Solar electricity may be some way off, solar heating need not be; indeed demonstration solar homes have been springing up throughout the world. Instead, the Hydro publication discusses the virtues of *electric* heating: it is pollution-free (unless you live near the generating plant), and convenient (provided all of the wider ramifications of the electric expansion program are ignored). In anticipation of argument that other alternatives are socially preferable, Hydro claims the moral immunity of the market by invoking consumer sovereignty ('public demand,' 'industrial need'). What this amounts to in reality, however, is the acceptance of consumer preferences which Hydro itself has helped to create.[37]

The obsession of any CI institution with efficiency and growth leads, as we have seen, to centralized, high technology institutions with dominant experts and passive clients. Hydro is no exception: small-scale power plants run, for example, by wind, water, or diesel engines, have obvious advantages – no fuel costs for the former two, the possibility of achieving high efficiencies through the local use of waste heat in the latter – yet they do not figure in Hydro's plans. Equally absent is any role for local generating plants based on biological, geothermal, and solar energy. Instead, Hydro has been actively phasing out small generation plants. And Hydro expansion plans call instead for large nuclear power plants, with all their attendant social consequences, their hazards, and heat wastage.[38]

The active expert–passive client syndrome manifests itself here too, in the increasing removal of decision-making control from the public domain, a process well illustrated in recent efforts by Hydro at explaining its evaluation of the social consequences of future power developments.[39] The most crucial issues are simply ignored; there is no mention of the social control of electricity production and electric power planning, of the choice and location of power plants, nor are the negative physical and economic aspects of centralizing the energy system discussed. Value judgments behind decisions are rather summarily dismissed: 'To varying degrees external costs and benefits associated with the operation of Ontario Hydro are incapable of quantification, and therefore difficult to include in any decision-making evaluation.' But, of course, ultimately one always decides on the basis of unquantified information. The importance of unquantified, 'ultimate' considerations backing all quantified judgments is tacitly admitted elsewhere in the report, where we read: 'While economic differences can be quantified [*sic*!] many other differences cannot. As a result the trade-off process requires considerable judgement. *Control of this judgement is obtained by corporate review of the proposals* before major alternatives are committed for design and construction' (our emphasis). So, in characteristic CI fashion, the socially important decisions are to be matters for internal corporate judgment – judgment, pretty clearly, with a bias towards 'quantifiable economic factors.' What the process of corporate judgment amounts to we are unfortunately not told.[40] We would rather have it replaced by public choice, institutionalized in a public, critical process.

Lest one suppose that the present situation is really in the best public interest and could not drift any further towards authoritarian control, we would point out that already there is serious (if 'unofficial') talk of suppressing public discussion of Hydro's future borrowing requirements and their potential social consequences, because of the effect that this might have on borrowing prospects. Similarly, there is serious discussion of suppressing publicity concerning the safety weaknesses of the nuclear system, for fear that would-be plutonium thieves and saboteurs would gain an advantage. Both of these moves would of course be made 'in the public interest.' It should be disturbing to realize that, as the day nears when we find ourselves heavily dependent upon such an energy system, this justification for the removal of public responsibility will appear more and more appropriate. Thus do our massive, centralized CI institutions create their own conditions for removing themselves from public participation and control (and once again the passivity of the public becomes a self-fulfilling prophecy).

In the same spirit as in our discussion of alternative structures to the HCDS, we now point at some energy policy alternatives that try to avoid our present difficulties. The abstract motivation for the recommendations that follow is, once again, our insistence that CI institutions such as Hydro be transformed into match (M) institutions. This transformation can obviously not be brought about by simply opting for different ('alternative') energy-generating technologies. Rather, it requires a concerted intellectual effort and determined political action on the part of all citizens. If we nevertheless concentrate here on concrete technological and institutional changes, it is merely to demonstrate that the range of feasible alternatives is not as limited as one is sometimes led to believe. For a more detailed argument concerning the relationship between technological change and political change the reader is referred elsewhere.[41]

Current energy policy is characterized by its emphasis upon centralized, technologically massive, capital-intensive systems of high-grade energy production and distribution. We would recommend instead moving towards a decentralized, low-technology, low-environmental impact, locally adaptive energy system using a wide mix of energy production and distribution technologies. In particular: (1) the majority of low-grade energy (heating) should be supplied by a combination of solar, biological, and municipal waste generation; (2) small demands for high-grade energy should be satisfied through a combination of wind, biological, and small conventional power plants (hydro, fossil fuel); and (3) the demand for large quantities of high-grade energy (now greatly reduced) to be satisfied with specifically designed conventional power stations (hydro, fossil fuel) for the near-to-middle term future (15–25 years), the longer-term objective being to utilize a combination of non-conventional modifications of conventional technology (for example, tar sands oil plus hydrogen energy transmission system) and more sophisticated solar, wind, and biological technology on a larger scale. This generalized 'conserver society' alternative is rounded out with the institution of a broad class of regulations governing the reduction of energy wastage (for example, through regulations for housing design, industrial processes, waste disposal, transportation, etc.).[42]

In keeping with this general policy, we recommend the radical decentralization of Ontario Hydro through the formation of strong regional policy-making bodies comprising a mix of the regional representatives of provincial ministries and corporations (for example, Ontario Hydro, Ministries of Energy, Environment, Urban and Regional Development) and local municipal, business, and consumer interests. The primary policy-making concern of the regional bodies would be to determine those particular mixes in the energy production and delivery system which are most appropriate to the needs of their regions, taking into account a humane land use policy, idiosyncrasies of local lifestyle, and so on. One can, of course, anticipate strong variations of interest within these groups (for example, as between residential conservationists and those business interests which traditionally have stood to gain from Ontario Hydro's present policies), but we believe that the ensuing process of debate and compromise is an important part of the human political process. Indeed, we should want to insist that much of the proceedings of these regional policy-making boards should be publicly televised, and the boards themselves open to public intervention by groups not directly represented on them. Moreover, there should be a definite public process of impact assessment and critical review for each proposed policy.

The combination of regionalization and a diverse mix of energy technologies opens the way to a much greater variety and local adaptability of design. Not only do we recommend that regional boards make maximal use of this adaptability, but that much of the development of energy production and transmission technology and servicing be handed over to regional industry (working, of course, within the guidelines of policy laid down). In this way we would hope that much of the development, construction, maintenance, and regulatory bureaucracy which swells the sizes of current government institutions might be substantially decreased. At the same time, greater regional industrial activity can be promoted.

We anticipate the regional energy boards electing representatives to a provincial energy policy-making board of roughly the same composition. The primary purpose of this latter board would be to ensure the coherence and provincial 'balance' of the regional energy policies when viewed collectively. This would be done, not through a dictatorial decision 'from above,' but through a dialogue between the provincial and regional bodies concerned. Once again, we recommend that these processes be well defined and open to public scrutiny and participation. On the other hand, since much of the responsibility for the execution of energy policy can now be devolved upon the regions, the need for a massive centralized bureaucracy can be expected to decline. It is our contention that such a process of devolution of power and responsibility from the higher echelons of government and of large corporations to the regions and the communities constitutes our only chance to halt the dominating historical trend towards political and technological estrangement in our society.

COMPARISON OF THE HCDS AND ONTARIO HYDRO AS CI INSTITUTIONS

We have argued that both the Canadian health care delivery system and
Ontario Hydro are typical examples of what we have labelled commodity intervention (CI) institutions. Despite their greatly different social
roles there are broad similarities between the two, and many of these are
characteristic of CI institutions. To recapitulate: both institutions strive
to maximize the quantity of the commodity they deliver and the efficiency with which they deliver it. They seek to achieve this by relying on
an increasingly centralized, massive institutional structure. In the process both institutions have developed a technological conception of their
commodity which publicly discounts the social-evaluative dimension to
their mandates, while implicitly claiming for themselves the right to
build particular evaluations into their technological designs.

Both the medical institution and Ontario Hydro attempt to solve what
are essentially non-technical problems by a heavy application of sophisticated technology. The selection of new tools is in both instances often
uncritical and unimaginative. Both institutions are also to a considerable extent closed systems, fending off public involvement in their
decision-making processes by appealing to the technical character of the
commodity production process. Correspondingly, both institutions rely
on active expert–passive client relationships thereby discouraging self-
sufficiency. The result, in both cases, is a heavy emphasis on curative,
rather than preventive policies, an emphasis justified by an appeal to
'consumer sovereignty.' (We note, however, that these generalizations
do not fit many individual people working in these instutitions, and that
there are public pressures placed on health care and energy professionals which are in part responsible for the existing roles and which only the
pursuit of the 'public education' component of our alternative policy
could hope to alter.)

Both institutions are centralized, not only in their policy-making
processes, but in their physical locations as well. They assume that the
provisions of their services in the market has no repercussions upon the
character and quality of our social lives in general. They also assume
that the nature of the human roles in their institutions is justified by the
maximally efficient production of their commodities (compare the
human impact of an obstetrics ward nursing role with that of a nuclear
generator maintenance officer).

Of course, there are also differences between the two institutions. The
health care delivery system involves individuals somewhat more obvi-

ously and directly in moral decisions concerning their fellow humans; but the differences are less striking than the similarities. After all, an appropriate pattern of energy supply, like health, is fundamental to all living processes and so decisions in the realm of energy have widespread social ramifications. Our society's health and energy policies, along with our policies in a number of other fields (education, food production, communication, housing, transportation, etc.) determine in a profound fashion the pattern and quality of our social lives. The implication of our arguments is that the problems which our society faces in all of these fields are fundamentally very similar. Indeed, the majority of them can be traced back to two closely related circumstances: many of our key institutions have an inappropriate structure, and they rely on inappropriate forms of technology. The first circumstance expresses itself, as we have attempted to show for health care and electricity supply, in a narrow role conception and a fragmented and elitist approach to social functioning. The second point is illustrated by the reliance in all the fields mentioned above on technological procedures which tend to inhibit rather than enhance meaningful work roles and societal learning processes.

The two points are closely related: in our society the development and implementation of new technological tools is almost completely under corporate control, and (expected) profitability is often not a very good criterion when it comes to choosing appropriate types of technology. On the other hand, the fact that the structure of our institutions is inappropriate must to a considerable extent be attributed to market ideology. Within the market paradigm any conflicts of interest which might arise as a result of the development of controversial technologies will be settled by the market mechanism, and consumer sovereignty 'guarantees' that everyone can obtain the kind of technology he wants (provided he can afford to pay for it).

In the real world of monopoly and collusion, of market breakdown and unpriced externalities, however, the consumer is no longer sovereign and conflicts of interest remain unsettled, unless the government takes action. Since governments in market democracies (such as Canada) are torn between allegiance to both the corporate elite and the rest of the citizens, most such conflicts remain unresolved (at least in the longer term) and, indeed, they are aggravated to the extent that major decisions are taken without prior public debate.

There is in our society, not surprisingly, a strong tendency to deny the existence of fundamental conflicts of interest, or at least to ignore them.

And the structure of the CI institution, with its hierarchical ordering and lack of feedback, is eminently capable of doing just that. Our socio-environmental problems are therefore not solely attributable to a wrong choice of institutional goals or technological tools. Indeed, it would be futile to suggest alternative aims for our societal institutions (be it to pursue the ideal of a conserver society, to strive for the 'humane' use of technology, or whatever), without indicating in which political framework these aims – and other, possibly conflicting, ones – are to be discussed. It is clear that the structure and ideology of the typical CI institution preclude any meaningful debate on such issues. Match institutions, based as they are on structures which promote decentralized, iterative decision-making and conflict resolution, may prove to constitute the proper framework for shaping our future environment according to the developing conception of our societal goals and values.

NOTES

1 Our views are developed in some detail in our 1977 report to the Royal Commission on Electric Power Planning: C.A. Hooker and R. van Hulst, 'A Preliminary Study of the Conceptual and Institutional Structure of Energy Policy Making in Ontario and Its Policy Alternatives' – available on request from the RCEPP. The application of the ideas therein, however, is restricted to the energy field; the more general analysis and application is offered in Hooker and van Hulst, 'Systems and Value: The Systematic Structure of Socio-Environmental Problems' (to be published), and Hooker, 'Cultural Form, Social Institution, Physical System: Remarks toward a Systematic Theory,' in M.F. Mohtadi, ed., *Man and His Environment*, II (London 1976).
2 On this latter notion, see, for example, M. Hollis and E. Nell, *Rational Economic Man* (London 1975), and W. Leiss, *The Limits to Satisfaction* (Toronto 1976). Our views have been moulded especially by K. Polanyi, *The Great Transformation* (Boston 1944), and by the works of Robert Theobald, especially his *An Alternative Future for America*, II (Chicago 1972), E.F. Schumacher, *Small Is Beautiful* (New York 1973), and Sir G. Vickers, *Freedom in a Rocking Boat* (London 1970).
3 For more on these ideas, see notes 1 and 2 above and G. Stewart and C. Starrs, *Gone To-day, Here To-morrow*, and F. Thayer, *Participation and Liberal Democratic Government*, both prepared for the Ontario Committee on Government Productivity (Toronto 1972). See also Thayer, *An End to Hierarchy, an End to Competition: Planning and Organizing the Politics and Economics of Survival* (New York 1973).
4 C.E. Lalonde, *A New Perspective on the Health of Canadians* (Ottawa 1975)
5 N. Crichton, *Community Health Centres: Health Care Organization of the Future?* (Ottawa 1973); Science Council of Canada, *Science for Health Services, Report 22* (Ottawa 1974); Planning Committee for the Health Care System in Ontario, Chairman J. Fraser Mustard, *Report for the Minister of Health and the Cabinet Committee on Social Development: The Report of the Health Planning Task Force* (Toronto 1974) –

hereafter, *The Mustard Report*; H.R. Robertson, *Health Care in Canada: A Commentary*, Science Council of Canada, background study no 29 (Ottawa 1973)

6 J.M. LeClair, 'Recent Developments in Health Care Services in Canada: The Federal Point of View,' address to the American College of Hospital Administrators, Montreal, 20 Nov. 1972; and Science Council, *Science for Health Services, Report 22*

7 B.J. Kalish, 'Of Half Gods and Morals: Aesculapean Authority,' *Canadian Nurse*, 71 (June 1975), 30–6

8 *The Mustard Report*; H. Grafftey, *The Senseless Sacrifice: A Black Paper on Medicine* (Toronto 1972); H.A.P. Finnegan and E.J. Monkman (members of SHOUT), 'Attitudes to Health Care: Student Research in a Downtown Core,' *Canadian Family Physician* (Jan. 1972), 94–102; D.A. Pickering, 'Report of the Special Study Regarding the Medical Profession in Ontario,' a report to the Ontario Medical Association, April 1973

9 E. Vayda, 'A Comparison of Surgical Rates in Canada and in England and Wales,' *New England Journal of Medicine*, 289 (Dec. 1973), 1224–9; C.E. Lewis, 'Variations in the Incidents of Surgery,' *ibid.*, 281 (Oct. 1969), 880–4

10 G.H. Lowrey, 'The Problem of Hospital Accidents to Children,' *Pediatrics*, 32 (Dec. 1963), 1064–8; J.T. McLamb and R.R. Huntly, 'The Hazards of Hospitalization,' *Southern Medical Journal*, 60 (May 1967), 469–72

11 R.H. Moser, *Diseases of Medical Progress: A Study of Iatrogenic Disease* (Springfield, Ill. 1969); A.L. Cochran, *Effectiveness and Efficiency: Random Reflections on Health Services* (London 1972); J. Powles, 'On the Limitations of Modern Medicine,' *Science, Medicine and Man*, 1 (1973), 1–30.

12 Treatment of victims of heart infarction in intensive care units has not been shown to give better results than simple home treatment: H.G. Mather *et al.*, 'Acute Myocardial Infarction: Home Treatment and Hospital Treatment,' *British Medical Journal*, 193 (1971), 34–8.

For breast cancer the five-year survival rate is 50 per cent regardless of the frequency of medical check-ups or treatment. As a result of programs to encourage early detection a higher proportion of cases has been treated in early stages, but no corresponding decline in mortality has been observed. See Breast Cancer Symposium, 'Points in the Practical Management of Breast Cancer,' *Journal of Breast Surgery*, 56 (1969), 782; K. Atkins *et al.*, 'Treatment of Early Breast Cancer: A Report after 10 Years of Clinical Trial,' *British Medical Journal*, 2 (1972), 423–9. For other forms of cancer, see R. Sutherland, *Cancer: The Significance of Delay* (London 1960); see also B. MacMahon, 'Cancer,' in D.W. Clark and MacMahon, eds., *Preventive Medicine* (Boston 1967), 417–37. For an up-to-date review of the environmental causes of cancer, see, for example, J. Fraumeni, ed., *High Risk Cancer* (New York 1974).

13 I. Illich, *Medical Nemesis: The Expropriation of Health* (Toronto 1974). We strongly disagree, however, with Illich's analysis of the ills of industrial medicine and with his suggested cure. See the excellent review by Vincente Navarro, 'Health and the Industrial Model,' *Social Policy* (Nov.–Dec. 1975), 59–64, for incisive and lucid comments on why Illich's analysis is misleading.

14 W.D. McKee, ed., *Environmental Problems in Medicine* (Springfield 1974)

15 For example, 80 per cent of all cancers are thought to have environmental causes – and can therefore in principle be prevented. See Department of Health and Social Security, *On the State of Public Health*, Annual Report of the Chief Medical Officer ... 1970 (London 1971), and Fraumeni, *High Risk Cancer*.

16 Or, as Foucault once put it: 'When a judgment cannot be made in terms of good or evil it is expressed in terms of normal and abnormal. And when this last distinction needs to be justified, one resorts to what is good or bad for the individual.' From an interview with M. Foucault appearing in Actuel, *C'est demain la veille* (Paris 1973), 38.

On the myth of consumer sovereignty and the evaluative choices underlying the choice of economic/social theories, see, for example, the references in note 1 and Hollis and Nell in note 2 as well as D. Pearce, 'Are Environmental Problems a Challenge to Economic Science?' *Ethics in Science and Medicine*, 2 (1975), 79; J. Robinson, *Economic Philosophy* (London 1964); and L. Tribe, 'Technology Assessment and the Fourth Discontinuity,' *Southern California Law Review*, 46 (1973), 617–60.

17 L.D. Dickey, ed., *Clinical Ecology* (Springfield 1975)
18 See, for example, R. Alford, *Health Care Politics* (Chicago 1974); R. Dubos, *Man Adapting* (New Haven, Conn. 1965), and *Mirage of Health: Utopias, Progress and Biological Change* (New York 1971); and P. Sedgwick, 'Illness – Mental and Otherwise,' *Hastings Center Studies*, I, 3 (1973), 19–40. The latter is an excellent paper, one to which we are much indebted. We recommend Hastings Center Studies to all those concerned with these issues. See also the references in notes 14 and 17.
19 *Ibid.*
20 Clark and MacMahon, *Preventive Medicine*
21 How successful medicine has been in its attempts to 'objectify' the notions of health and disease, aided and abetted by the eager desire of social studies to become recognized as 'sciences,' can be seen in the extensive literature on medical sociology where illness is characterized as a problem in social control or a form of social deviance. Some of the most influential works are T. Parsons, *The Social System* (Glencoe, Ill. 1951); E. Friedson, *Patients' Views of Medical Practice* (New York 1961); and B. Mechanic, *Medical Sociology* (New York 1968)
22 For a proposed 'Faustian bargain' in the nuclear energy field, see A. Weinberg, 'Social Institutions and Nuclear Energy,' *Science*, 177 (7 July 1972), 27–34. For the Port Hope problem, see, for example, the *Globe and Mail*, Toronto, 13 Jan. 1976 – 'Ontario Won't Divulge Radiation Readings Found in Port Hope Tests'. Radiation readings in several Port Hope homes and in one school exceeded 50 pCi/litre (the accepted international standard *for occupational exposure* is 30 pCi/litre, with an induced annual lung cancer risk of 6 ± 3 cases per year per 10^6 miners). Translation of hard-to-grasp radiation readings into estimated lung cancer risks was carefully avoided, however, and the Ontario health minister, Frank Miller, went on record saying: 'We don't have a health problem, we have a scare problem.' For recent literature on radon induced lung cancer risks, see W. Jacobi, 'Problems Concerning the Recommendation of a Maximum Permissible Inhalation Intake of Short-Lived Radon Daughters,' in E. Bujdosó, ed., *Health Physics Problems of Internal Contamination* (Budapest 1973).
23 To judge from the growing literature on organizational issues in medicine, an increasing number of administrators think, for example, that the 'production parameters' of diagnoses and cures can be so exhaustively known that computers can take over the task and that the major criteria for the organization of hospital roles should copy the time-efficiency models of large industrial enterprises. See, for example, many of the papers in *Technology and Health Care Systems in the 1980's*, National Center for Health Services Research and Development, National Technical Information Service, US Department of Commerce, 1973. See also H.I. Greenfield, *Hospital Efficiency and*

Public Policy, Center for Policy Research (New York 1973); C. Perrow, 'Hospitals: Technology, Structure and Goals,' in J.G. March, ed., *Handbook of Organizations* (Chicago 1965); R.M. Bailey, 'Economies of Scale in Medical Practice,' in H. Klarman, ed., *Empirical Studies in Health Economics* (Baltimore 1971); S. Greenberg, *The Quality of Mercy: A Report on the Critical Condition of Hospital and Medical Care in America* (New York 1971). As elsewhere in these notes, this is only a representative list from a much larger literature. Not surprisingly, this managerial trend is becoming increasingly seriously disputed. See, for instance, M. Michaelson, 'The Coming Medical War,' *New York Review of Books* (1 July 1971), and B. Mendelson, J. Wazey, and I. Taviss, eds., *Human Aspects of Biomedical Engineering* (Cambridge Mass. 1971).

24 In December 1975 the Ontario health minister, Frank Miller, announced that the government would attempt to reduce hospital spending by $50 million by eliminating 3000 hospital beds and reducing staffs by 5000 people. The reductions were subsequently accomplished mainly by closing several small hospitals. 'Improved highways and good ambulance systems enable the closing of small hospitals without loss of health care quality,' the minister maintained. *Western News*, London, 21 Oct. 1976

25 See above, note 9

26 See above, note 5

27 See, for example, R.J. Ledogar, *Hunger and Profits: US Food and Drug Multinationals in Latin America and the Carribean* (New York 1975); C. Levinson, *The Multi National Pharmaceutical Industry*, International Federation of Chemical and General Workers Unions (New York 1972); M.H. Sjöström and R. Millson, *Thalidomide and the Power of the Drug Companies* (London 1972); A. Class, *There's Gold in Them Thar Pills: An Inquiry into the Medical-Industrial Complex* (London 1975).

Contra-indications for drugs are frequently underplayed and doctors are often subjected to an arsenal of hard-sell techniques. Even after a known medical disaster, such as the massive misuse of the drug chloramphenicol, the profession is often unable to alter the prescription habits of its own physicians. See T. Bodenheimer *et al.*, *Billions for Band-Aids* (San Francisco 1972)

28 See R.H. Hall, *Food for Naught: The Decline in Nutrition* (Hagerstown, Md. 1974). For an excellent discussion of the social implications of modern food technology, see Magnus Pyke, *Technological Eating or Where Does the Fish Finger Point?* (London 1972).

29 See, for example, M. Foucault, *The Birth of the Clinic: An Archaeology of Medical Perception* (New York 1972), and note 18 references

30 See above, notes 4 and 5

31 For information on Ontario Hydro, see Ontario Hydro, *Long-Range Planning of the Electric Power System*, Report no 556SP (Toronto, 1 Feb. 1974), and for information on Ontario energy use, see Advisory Committee on Energy, *Energy in Ontario: The Outlook and Policy Implications*, II (Toronto 1973) and the subsequent annual up-dates of the same title by the Ministry of Energy. For Hydro's legislative mandate and responsibilities, see Task Force Hydro, *Hydro in Ontario: A Future Role and Place*, Report no 1 (Toronto 1972) and Ontario, *The Power Corporation Act* (Toronto, Sept. 1975). The two quotes in the text are, respectively, from the first and last of these documents.

32 Ontario Hydro, *Submission to the Ontario Energy Board – System Expansion Program*, I–IV, (Toronto, 19 Dec. 1973)

33 See, for example, Ontario Energy Board, *Ontario Hydro: Bulk Power Rates for 1976*,

Parts I and II, (Toronto, 10 Oct. 1975). Ontario Hydro, Rate Review Department, *Correspondence Pertaining to Public Hearings into Ontario Hydro's Long Range Planning Concepts* (Toronto, March 1975)

34 Since we wrote this paper in early 1976 the degree of inflexible attachment to the 'ever bigger,' 'technical fix' syndrome has been well illustrated by Hydro's performance before cabinet and the parade of *ad hoc* government bodies investigating Hydro which have been created over the past few years. Even in the face of lowered load-growth forecasts Hydro has clung to its massive expansion program. It has already been ordered by the provincial treasurer to cut back on borrowing because of the strain placed on provincial borrowing credibility (1976). Only after mounting government and public pressure was it persuaded to modify its program even marginally. Evidence is now emerging that not only did Hydro deliberately plan an expansion of nuclear heavy water facilities which would have been wholly provincially utilized only in the most optimistic expansion scenario, but it rushed itself and the province into this commitment well ahead of need, or even of proper planning, in order to beat Hydro-Québec to any potential export market. The Royal Commission on Electric Power Planning has been forced to delay its interim report on nuclear energy policy because Hydro and other government agencies in the nuclear field saw fit to withhold evidence from it of secret meetings on nuclear safety. Indeed, the very parade of committees, commissions, and the like investigating Hydro in the past decade is eloquent testimony to increasingly desperate government efforts to obtain some sort of grip on Hydro's activities. The most recent select committee of the legislature has been forced to use its powers of subpoena because Hydro personnel otherwise will not appear before it; for reports on this see the *Globe and Mail*, Toronto, 13 June and 20 July 1978, and the proceedings of the select committee.

35 See Hooker and van Hulst, note 1 above

36 See, for example, Claude M. Summers, 'The Conversion of Energy,' *Scientific American*, 224, no 3 (Sept. 1971), 148–63. For the distinction between high-grade and low-grade energy, see William D. Metz, 'Energy Conservation: Better Living through Thermodynamics,' *Science*, 188 (23 May 1975), 820–1; and the reports of the American Physical Society Summer Studies cited there.

   On alternative energy technologies, see, for example, Task Force on Energy Research and Development, *Exploit Renewable Resources: Solar Energy*, R & D Program Outline (Ottawa, March 1975); Peter Middleton and Associates, *Canada's Renewable Energy Resources: An Assessment of Potential* (Toronto 1976); A. Lovins, *Soft Energy Paths* (Cambridge, Mass. 1977), and his 'Renewable Energy Technology: The Road Not Taken,' *Foreign Affairs*, April 1975.

37 Electric heating is also described as highly efficient, and it is – so long as the inefficiencies of generating it in the first place are ignored; cf. the references of notes 39, 40. Some of the wider social ramifications are discussed further in the essays cited in notes 1 and 39.

   The main emphasis of Hydro has not been so much on cheaply providing the small amounts of electrical power required by households but rather on satisfying the large industrial customers who account collectively for roughly 40 per cent of the provincial electricity consumption. Through the low cost of electrical power resulting from the strongly regressive rate structure, industrial energy use has become exceedingly wasteful. A rather similar situation prevails in the commercial sector (the wastefulness of the lighting, heating, and cooling systems of large office buildings is notorious).

Hydro's (largely token) propaganda for energy conservation seems to be directed mainly at the average Ontario household. This should be contrasted with the fact that the industrial and commercial sectors together account for roughly 70 per cent of the provincial electricity consumption. Moreover, the tendency of this propaganda has been to blame the individual's use of electrical gimmicks for the present energy problems. But although the intellects and wills of the populace have surely to be engaged on this problem, it is not electric toothbrushes and colour televisions that call for crash programs in nuclear plant construction. (Home appliances are estimated to account for one per cent of Canadian energy demand.) Opportunities for energy conservation in the industrial and commercial sectors are far more relevant than those centred around the household. The same applies to the substitution of sources of low-grade heat for high-grade electricity.

38 The amounts of waste heat from large nuclear stations will probably prove too large and too localized to be effectively utilized. While Hydro has until recently excluded small generating plants from its plans, as a result of government pressure (including a letter from then minister of energy James Taylor) it is now considering a modest reintroduction of smaller hydroelectric plants. The outcome remains to be seen. See the *Globe and Mail*, 20 July 1978

39 Ontario Hydro, *Socio-Economic Factors*, submission to the RCEPP (Toronto 1976). Both quotations in the text following are from this document.

40 Cf. note 34

41 For two excellent accounts of the complex social relations between technologies, institutions, and social goals and values in this context, see J. Habermas, 'Technology and Science as "Ideology,"' in *Toward a Rational Society* (Boston 1974), and David Dickson, *Alternative Technology and the Politics of Technical Change* (Glasgow 1974); see also our essay, note 1 above.

42 Our views have been elaborated in some more detail in the report cited in note 1, although it too is perforce only a preliminary statement so far as detailed application is concerned. For supporting material, see *Canada as a Conserver Society* (Ottawa 1977) by the Science Council of Canada and the council's *Conserver Society Notes*, now published by *Alternatives* magazine (Trent University). Both of these periodicals are highly recommended sources. See also James Ridgeway and Bettina Conner, *New Energy: Understanding the Crisis and a Guide to an Alternative Energy System* (Boston 1975); David B. Brooks, *Energy Conservation and the Roles of Government,* Department of Energy, Mines and Resources (Ottawa 1976); *Scenarios for Canada to 2025* (Ottawa 1977). See also E.P. Gyftopoulos, L.J. Lazaridis, and T. Widmer, *Potential Fuel Effectiveness in Industry* (Cambridge, Mass. 1975). A striking example of the potential savings is offered by the conclusion reached here that the paper industry (one of the largest users in Canada) could be very nearly energy self-sufficient because of the heating value of its waste products. P.F. Chapman, 'Energy Costs of Products and Fuel Conservation,' *Energy Policy*, in press; C.A. Berg, 'Conservation in Industry,' *Science* (19 April 1974), 264–70

8
The nuclear power controversy

R. van Hulst

The Canadian government foresees a gradual shift from the present energy supply system based largely on fossil fuels to an electric society, with predominantly nuclear electricity supplying 90 per cent of the country's energy.[1] The plans to rely increasingly on nuclear electricity are usually defended by pointing to the technical and economic advantages of the nuclear option, especially in a time when foreign sources of oil are perceived as unreliable and domestic supplies, at least the more accessible ones, are starting to dry up.

Yet, nuclear power also presents many problems. The use of this form of energy is well known to involve hazards of a particularly serious nature. Radioactive compounds created in the course of nuclear reactions are potent poisons, much more potent and insidious than chemical poisons. The ability of nuclear engineers to guarantee the extremely high degrees of containment necessary to avert disaster became controversial at an early stage in the history of civilian nuclear power. Reactor accidents due to carelessness, unjustifiably poor working conditions in uranium mines, and an attitude of blind technological optimism all combined to call the claims of the proponents of nuclear power into question.[2]

The customary economical-technological justification of nuclear power was further undermined when the anti-nuclear movement rediscovered the problem of externalities: if the economic mechanism is incapable of accounting for certain impacts of nuclear power generation, and if economic and technological criteria are the principal or even the only ones used, then such 'wider societal impacts' will be left systematically out of consideration. Yet, opponents of nuclear power argued, such impacts are of considerable importance. Health risks, the local and community impacts due to large nuclear plants, curtailment of civil liberties necessitated by the weapons potential of nuclear fuel and wastes, all these are costs that are borne by segments of the population and do not find expression in the economic criteria used to assess the desirability of nuclear projects.[3]

Finally, problems created by the long-lived radio-isotopes in nuclear wastes drew attention to the moral implications of nuclear programs. Is it morally justifiable that we oblige the next 10,000 generations of humanity to guard the nuclear wastes that result from the desire of this generation to generate large amounts of electric energy?[4] Can we impose considerable radiation risks on the world's population for the next 80,000 years because we want to mine large quantities of uranium and cannot be bothered to clean up the mine tailings?[5]

All that considerations of this kind do, however, is to emphasize the political nature of the nuclear controversy. For when governments do go ahead with nuclear programs the decision to do so is rarely based on a careful public consideration of these issues. On the contrary, the decision of whether or not to adopt nuclear power generation is seen as essentially a technical matter, the responsibility of a small group of experts. The recent concession by governments and public utilities to allow some form of public input does not in the least change the fact that the key decisions, such as what energy technologies to develop or how much electricity to generate and in what way, continue to be made by a small economic and technological elite.[6] It is often argued that technologists and economists are, through their professional preoccupation with a narrow set of criteria, the groups least qualified to decide on morally controversial issues.[7] But this, of course, is nonsense: by no means every technologist or economist is so preoccupied, and, besides, technologists and economists are surely just as entitled to have their own moral values as anybody else (even to decide whether something is a moral issue is itself a moral issue). The point to criticize is rather that a small minority tries to impose its convictions on the majority of citizens by concealing the morally controversial nature of its plans. It is of course a fact that moral attitudes are often influenced by perceived interests and that there are conflicting opinions about what is morally justified. Indeed, the possibility of conflicts of interest among different societal groups is not readily admitted in our culture: we rather like to think that what appears to be a conflict situation is really based on a misinterpretation of the facts or on mutual misunderstanding.

This essay is an attempt to demonstrate that the nuclear controversy is in essence a political controversy. The subject of the controversy is how our society is to guide its technological development, and not whether nuclear wastes can be safely stored for eons or whether the risk of catastrophic nuclear accidents can be effectively minimized.

The motives for developing civilian nuclear power technology were mainly ideological and economic ones, as was the case with most other major technological projects developed in modern societies. Nuclear power generation in a few large generating stations apparently fitted the rather vague and unarticulated ideals about how the supply of essential services ought to be organized. It was also expected to lead to higher eventual pay-offs to its developers than competing technologies (nuclear reactors were primarily developed to power military submarines; the modifications required for civilian power generation were minimal). This is not to say, of course, that the development of nuclear energy was a wholly rational and planned affair, for this can never be the case in technology research and development. But to the extent that choices were to be made the decisions were strongly influenced by ideological and profit motives, and not by considerations of moral righteousness.[8]

Neither is the choice whether or not to continue to build nuclear generating stations likely to be made solely on the grounds of moral criteria. Existing commitments and large investments on the part of certain groups in our society effectively preclude this. (To be sure the government could provide compensation to the groups involved, thus effectively buying off their right further to pursue the nuclear option. This would entail, however, putting a monetary price on anticipated future profits and, probably, doing this publicly – a step not likely to be welcomed by either government or private companies.)

Conflicts of interest such as the ones referred to above are not solved by moral appeals; nor will they disappear by themselves. Only the political process of public discussion and political action can effectively deal with such problems.

It is no accident that in our society scientists and technologists are not only in charge of developing technology but also of applying it to societal problems and, moreover, of justifying such applications. The fact that appeals to 'scientific evidence' to justify morally or politically controversial decisions are widely regarded as perfectly legitimate in our culture allows those acting in the name of science to ascribe to the existing order a historical necessity. Arguing against some feature of the existing order becomes equivalent to attacking science and exposes one to the accusation of being irrational, unscientific, unintelligent.

This brings us to the second point to be explored in this essay, namely the legitimating role of science with regard to technological change.

Science, or what goes under that name, is now rapidly replacing religion in the public mind as the final arbiter in moral matters. Thus a new privileged class is created, that of (applied) scientists and technocrats. And the ordinary citizen feels more disenfranchised than ever. There is certainly a growing dissatisfaction among Canadians over the way in which discussions which affect them crucially have been appropriated by small groups, whether they be government bureaucracies, big multinational companies, or scientific research teams.[9] Much of the dissatisfaction is diffuse, however, and expresses itself in feelings of alienation, of being at the mercy of the system.

The significance of the nuclear power controversy is that it provides a key example of how our society chooses its technological tools and subsequently justifies its choices. The issues in this controversy are of such importance and so conspicuous that a debate involving a large proportion of citizens could be an important learning experience. But if this is to happen the discussion must deal with the problem at a much more fundamental level; the arguments that have been presented to date have been rather non-committal and superficial.

THE NEED FOR ELECTRICITY

Nuclear power generation is advocated by many electric utilities and by governments on the double grounds that an increased demand for electricity can be expected in the future and that nuclear power generation provides the cheapest and technologically best developed way to produce this electricity. In this section the first part of the justification, the prediction of an increased demand for electricity, is examined. The way in which the choice of nuclear power is justified will be the topic of the next section.

As we know, the demand for electricity has been rising in Canada at a steady annual rate of about 7 per cent.[10] To predict load growth for the immediate future, utilities take into account expected changes in the number of customers, in the economy at large, and, to a lesser extent, in the price of electricity (elasticity of electrical demand is usually supposed to be very low and substitution among energy forms is considered negligible in the short term).[11] Utilities have developed a certain expertise in such short-term load forecasting, and their predictions are rarely far from what actually occurs.

Long-term load forecasting, however, is quite another matter. The implicit assumption of *ceteris paribus*, of all else remaining the same, on which short-term load forecasting is clearly based, becomes difficult to maintain. Substitution among different forms of energy can no longer be ruled out, and entirely new ways of generating electricity may be developed (for example, small local solar and wind generators, methane burning fuel cells, etc.). Neither can the introduction of radical conservation programs be excluded: recent studies predict that electricity consumption can be reduced very considerably without appreciably affecting lifestyles by making use of technical knowledge already available. And, most important of all, it is well known that electricity consumption (and energy consumption in general) depends on a host of policy decisions regarding housing, transportation, regulation of industrial processes, regional planning, and so on. To 'predict' these, even with a sophisticated econometric computer model, is of course quite impossible, and the only thing the load forecaster can do is to assume that there will be no policy changes and no changes in societal priorities.

This, in fact, is precisely what long-term load forecasting amounts to: despite the aura of scientific thoroughness imparted by the use of computers and sophisticated models the future is predicted by simply extending past trends. Under the 'everything as usual' assumptions fed into the computer, the programs tend to predict, not unexpectedly, that electricity consumption will continue to grow at historical rates.[12] Since excess power tends to attract customers, and since utilities continually manipulate demand to make it correspond to production (based on past predictions), such predictions tend to become self-fulfilling prophecies.

One might argue that the prediction of future electricity demand does give rise to serious problems, due among other things to the societal significance of electricity, but that the general principle of consumer sovereignty on which it is based is nevertheless a laudable one. The assumption is, of course, that the market will ensure, through the price mechanism, that consumer demand will be met, provided that such is feasible and that consumers are willing to pay the price. Since meeting the demand for technologically sophisticated products usually requires some preparation, producers are forced to predict future demand, and it is in their own interests to do this as accurately as possible.

Let us examine this line of argument. First, even if one assumes that Canadian society is sufficiently close to the theoretical ideal of true market society (with free enterprise, a fully competitive market, and so on) so that the standard economic consequences apply, then there are still problems. The production system in Canada (as elsewhere) has come to be organized in increasingly large, centralized units. The increasingly long planning horizons that are necessarily involved in the production of virtually all commodities oblige producers to deal with predicted future demands and to invest considerable resources into making demands conform to corporate objectives. This removes us already quite far from the ideal of consumer sovereignty.

Furthermore, there exist many needs for the fulfilment of which we do not wish to rely solely on the market. The provision of medical care, education, police and armed forces, and many other needs perceived by society has been wholly or partially removed from the market. And even the generation and distribution of electricity, which in Canada is effectively a government monopoly, does not fit the market model. It is instructive to compare electric power planning with planning for the provision of other societal needs: we do not ask what the future demand will be for theatres, highways, national parks and nature reserves, museums, or armies. We expect governments to decide what our future 'needs' in these and many other fields will be, and which of these we can afford to satisfy.

Electrical load forecasting is in this light somewhat of an anachronism: in most societally sensitive fields the setting of future production goals is a matter of policy – not one of trying to estimate future demand. This circumstance is undoubtedly related to the fact that the electric utilities in Canada have a fundamentally schizoid role. On the one hand they are semi-government institutions that enjoy a legally recognized monopoly position and are subject to policy guidance by the government (notably with regard to pricing and the export of power). On the other hand, however, utilities are organized and behave in many ways as if they were market institutions. There are some indications now, particularly in Ontario, that the inappropriateness of this circumstance is starting to attract wider attention.[13]

Apart from the fact that the provision of key commodities is not left to be regulated by market forces in our society, the free market or free enterprise system is anyway a theoretical chimera: nothing even remotely like it did ever really exist. One of the effects of monopoly and collusion, both of which are widespread, is that only certain commodities will be produced, namely those that promise to be most profitable. This already restricts the range of choices open to the consumer; together with the restrictions imposed by government policy it relegates the dogma of consumer sovereignty to the domain of fiction.

Some further thought suggests that the focus on commodities is itself misleading. Our perceived needs, whatever their nature and origin, are clearly dependent on our wider societal and cultural conceptions of what constitutes a rich and satisfying life. The subordination of needs to commodities, however, makes us entirely dependent, for the definition as well as the fulfilment of our needs, on the narrow range of options that happen to be provided by the market or the government.[14]

To see how this confusion between needs and commodities has arisen let us take a closer look at the way in which the market deals with needs, in particular societal needs. We have already seen that in a market society[15] needs are assumed to be indicated by demand in the market (though ultimately demands are thought to derive from subjectively perceived 'wants'). Pressing our claims for satisfaction of wants on other members of society, or denying such claims, are of course moral actions. But the economic market obscures or ignores the morality by representing such actions only in terms of prices and 'economic rationality.' Economists often claim economic theory to be objective and they evidently mean by this that no individual moral judgments figure explicitly in its formulation. It is rarely realized, however, that the adoption of the entire free market system of exchange of goods and services represents a morally fundamental choice.[16] This is not the place to explore the extent to which this choice shapes our social fabric and to show that it is at the root of many of our present social and environmental problems.[17] Our society has chosen to represent the institutionalization of the economic market not as a historical moral choice but instead as a fact of nature. The economic myth[18] thus generated now serves to fix human social relations in a form that precludes the critical examination of their moral basis.

At this point it is useful to distinguish between contemporary demand and future demand. First consider contemporary demand. So long as we can safely assume that the basic structure of human wants (technically, preference orderings) changes substantially only over the medium to long term, we may effectively neglect the real psycho-dynamics of human wanting and assume as given a fixed structure of human wants (preference orderings). In this case we recover the framework of conventional market economics, which entirely excludes all considerations of human psychology, and we may say that changes in contemporary demand depend only on changes in price (to a degree measured by the price elasticity of the commodity in question) and on changes in the number and buying power of potential buyers. But this situation changes as soon as we consider future demands in the longer term, for then there is much greater pressure to admit that human preference orderings can undergo major shifts. Since the transformation of personal evaluations is a moral question, and the societal response to such changes a socio-political issue, at least in the longer-term perspective we are in turn forced to admit that social and political factors necessarily intrude into the otherwise neatly segregated picture of the economic system. The proper level at which to study social policy is normative political economy, not 'objective' economics.[19]

THE NEED FOR NUCLEAR POWER GENERATION

Once the need for producing some commodity has been established, how does market society go about choosing a production technology? In the case of electricity generation, how are generating technologies chosen? The answer, of course, is that the choice between competing technologies is once again made on supposedly 'objective' economic grounds. These typically take the form of a comparison of present economic costs, that is, costs in so far as they find an expression in market prices.[20] Externalities, or 'costs' that do not for some reason find expression in current market prices, are thereby immediately discounted. This problem is now widely recognized and economists have developed new tools that are supposed at least partially to overcome it.

Before discussing these techniques, however, mention must be made of an even more fundamental problem inherent in the use of economic cost effectiveness criteria as a basis for policy-making. It can roughly be characterized as follows: there is always a range of diverse future policies, any one of which can be made to appear economically optimal if only society pursues an appropriate historical course of action leading to its realization. Considerations of economic optimality can therefore provide little help in determining policy.

There is essentially a two-staged argument for this conclusion. The first stage consists in the observation that 'instantaneous' marginal economic analysis of any variety cannot rationally decide between historical courses of action. This is so because current price structure will always favour existing technologies over new technologies (unless the existing technology is literally falling apart), and thus the decision to extend the existing policy will always be favoured. And this will be the case even though in other historical circumstances it would have been an alternative policy that would have been favoured at the margin. The conclusion holds irrespective of the unpriced historical-social repercussions of following the current policy (or indeed of following any of the alternatives).

What is at issue here can best be seen with the help of an example. Small electric generators powered by the wind, falling water, or an internal combustion engine are said to be uneconomical sources of electricity in comparison with large fossil-fired or nuclear generating stations. But the reason that they are uneconomical lies precisely in the fact that our present electricity supply system, through its very reliance on centrally generated electricity, minimizes the market demand for small generators, thus preventing their mass production and keeping their price high.[21]

The second stage of the argument consists in pointing out that even the use of time-integrated marginal analysis of some sort is also of no avail in reaching an 'objective' position, because most, if not all, of the socially important resources involved will have their prices determined by political decision rather than by the free market (they will be 'policy-priced' rather than 'market-priced'). Thus the economic analyses serve only to reflect historical political commitments, commitments made on the full range of human evaluative grounds rather than on any 'objective' economic criteria. To see the force of this point one has only to reflect on current disputes about the pricing of Canadian oil, which is clearly 'policy-priced,'[22] and to remind oneself that government tax structures, tariffs, subsidies, etc., all constitute examples of policy pricing. (In any case, there is little evidence in the public domain that economists have moved beyond 'instantaneous' marginal analysis in their arguments.)

Partially as a reaction to the criticism of environmentalists that only narrowly economic criteria are employed in comparing alternative technologies, a number of economic techniques have been developed which supposedly overcome this objection. Known collectively as project evaluation techniques (or environmental impact assessment techniques), they consist of such economic tools as cost-benefit and matrix evaluation analysis.[23] That such techniques provide 'objective,' 'scientific' approaches which, when implemented in some mechanical fashion, allow one to demonstrate beyond any doubt the optimal course of action is, of course, a fiction. First, what to count as 'costs' and what as 'benefits,' and to what extent (especially to what *monetary* extent) is transparently an evaluative issue. Second, beyond this, the decision-producing methodologies themselves have their own underlying value content which itself needs to be critically evaluated.[24] If both of these sorts of evaluations are not made explicit, and *a fortiori* if the impression is given that no value judgments are made at all, the analysis only succeeds in obscuring the real societal issues.

The attitude towards technological choice described above is bolstered by a conception of technology best indicated as *technological determinism*, or the cargo conception of technology. Just as 'cargo' in the so-called 'cargo cults' commonly practised by certain 'primitive' tribes, technology is often viewed as a gift from heaven, perhaps evoked by the performance of some secret and poorly understood ritual. Both the cargo cult and technological determinism explicitly accept the current material values and seek to provide for their followers a means of achieving them.[25]

As long as technology is regarded as value-free and morally neutral, the choice between different technologies can safely be made the responsibility of a small group of specialists. But the way in which technology as an institution is organized – and has to be organized – dispels any doubt as to its value-ladenness: whatever the motives are that drive technological innovation (profit, domination, humanitarianism), it can hardly be denied that they involve value judgments. It is not surprising, therefore, that the technological products of a society reflect in many subtle ways the predominant world view.

The fact, for example, that many societies at the present time rely on strongly centralized energy supply systems and almost completely neglect locally available sources of renewable energy is no mere historical accident, but a direct consequence of both political factors and culturally determined preferences. Among the latter are, for example, the tendency in our culture to encourage far-going specialization and our culturally preferred form of control (through the supply of commodities rather than by making explicit appeals to authority).

On the other hand, one ought to be wary of lapsing into a simple-minded technological fetishism. Sometimes, for example, one gets the impression from opponents of nuclear programs that solar energy is regarded as capable of curing all our societal ills, as 'the healing balm of the late twentieth century.'[26] Nuclear power, of course, is condemned with as much fervour as solar power is praised. This, however, is just as misleading as representing technology as non-committal and obscuring the wider implications of its use. One simply cannot judge a technology in itself, separate from its societal environment. And neither does one have to: the choice is always between policies, not technologies.

Nuclear power is now chiefly advocated on the grounds of its alleged technical and economic superiority. The argument developed above, however, suggests that such considerations are often mere rationalizations for decisions taken on much wider grounds.[27] The history of civilian nuclear power illustrates this very clearly. Already the technology's military origins leave little doubt as to its original function, which was to power submarines and to utilize the existing stockpiles of enriched uranium. The subsequent development of the technology by industry (or, as in Canada, for industry by a crown corporation), its promotion by manufacturers of heavy equipment, by the electrical equipment industry, and by uranium mining companies, all suggest that corporate interests have always been an important driving force behind nuclear programs.[28]

Even a cursory examination of the present distribution of research and development efforts in the field of energy technology shows that the same forces continue to mould our energy policies. Best funded are those technologies that fit in with our present centralized energy supply system, require the most sophisticated engineering efforts, and continue to rely on ever-increasing quantities of scarce non-renewable resources. Even energy sources with a potential for local application using 'intermediate' technology are bent to fit the ideological and financial interests of the technological planners: solar energy research, for example, is concentrated in the aerospace industry and principally directed towards electricity generation in large 'solar farms.' The infinitely more simple and less wasteful possibility of direct solar space heating, employing locally built solar collectors, is hardly given any attention. And so we continue to develop energy supply systems that promise to deliver all the energy we need (provided we are willing to pay the price), but do not leave us the freedom to determine how much we will need and to satisfy our needs in a manner that reflects our personal values.

THE NUCLEAR POWER CONTROVERSY

Civilian nuclear power was announced two decades ago as a clean and abundant source of cheap energy. Since then the threat of illicit use of reactor plutonium and of the proliferation of atomic weapons has become a source of great concern.[29] We have also learned considerably more about the potential health effects of relatively low levels of radiation,[30] and we now realize that uranium reserves are quite limited.[31] Uranium prices have recently quadrupled and there has been a sharp rise in the construction costs of nuclear power plants. As a result, hardly anybody now dares to advocate nuclear power programs in such elated terms any more. Arguments that appeal to the economic and technological superiority of nuclear power generation still predominate, but the accent is now on the alleged lack of feasible alternatives. And even proponents of the adoption of nuclear power admit that there exist some quite serious problems. Most widely discused are the risk of serious accidents, claimed to be extremely small,[32] and the risk of illicit weapon fabrication out of nuclear wastes. Although the potential consequences are frightful, advocates of nuclear power believe that the risks are well worth taking in view of the considerable benefits that the technology would engender.[33]

Others see problems sufficiently serious to call for an immediate halt to nuclear programs.[34] Their concerns centre around the health hazards and accident risks of nuclear power generation, the weapons potential, and the long-term hazards of nuclear wastes. More particularly, disagreement arises from two sets of circumstances. First there is a growing concern that in their zeal to represent nuclear power as an attractive option, advocates of rapid nuclear development have persistently ignored some of the problems that surround the technology. This lack of faith in the assurances of the experts that risks and hazards are minimal is, after all, quite understandable if one examines the record of overly optimistic predictions and misleading statements that have issued from the nuclear industry and from some government departments. Allowable exposure limits to radiation have been reduced by several orders of magnitude in the last two decades, and they are still criticized as too lenient.[35] Worrisome statistics, such as the estimated effect in terms of excess cancer deaths of normal emissions from nuclear power plants, and the potential results of a large accident have been surpressed,[36] and the police-state measures that are clearly necessary to prevent diversion of nuclear fuel and wastes have persistently been ignored. The potential benefits of nuclear power generation, on the other hand, are blown quite out of proportion by inflating expected future energy demand and by claiming that nuclear power will be capable of meeting a large part of this demand.[37]

The ways in which the true hazards of nuclear power generation are underplayed, particularly in statements produced for mass consumption, are often exceedingly subtle and ingenious.[38] One of the devices employed is to suggest, by the use of linguistic analogies, a non-existent familiarity with and control over nuclear problems. When radioactive materials have been accidentally spilled, for example, the site of the spill and the people exposed need to be 'decontaminated,' as if some vermin or germ had spread which can easily and completely be destroyed. But only the passage of time will reduce radiation levels and there exists no feasible technology for extracting all or even nearly all of the radioactive material from the environment or from a person's body once it has been sufficiently diluted. 'Decontamination' in such cases is therefore utterly impossible and the frequent use of the term is simply a case of professional mystification.

Similarly misleading is the customary comparison of the estimated risk of a serious nuclear accident with the calculated odds of dying in a plane crash or a car accident. To see this it is sufficient to realize that any risk analysis of *new* technology incorporates many guesses and, more importantly, trade-offs and value judgments which enter the analysis, for example, on the many occasions where it is necessary to set maximum allowable levels of radiation exposure, to choose between competing assumptions (for instance, highly conservative ones or 'realistic' ones). To compare the risk of a severe nuclear plant accident with more conventional risks is therefore highly misleading. In the latter case, risk estimates are based on past occurrences of accidents, whereas in the former a long and tortuous analysis requiring many *ad hoc* assumptions was necessary to arrive at a superficially comparable statistic. The situation has very appropriately been compared with that of manned spaceflight: nobody has yet lost his life going to the moon, and a risk analysis comparable to the nuclear one would probably describe the chances of a serious accident as being extremely small; yet few people would be prepared to take the risk.

One could continue listing such misleading and highly unscientific statements, which nevertheless draw on the authority of science for their legitimation. One more example should suffice. It is often suggested that very low levels of radiation exposure are completely safe (the slogan is 'less radiation than the amount which causes a sunburn').[39] Virtually all the evidence, however, supports the thesis that there is no 'lower limit' below which radiation is harmless, but that there is a linear relationship between dose rate and cell damage.[40] And any junior high school student knows that sunburn is caused by ultraviolet rays, not by strongly ionizing radiation. The process of setting maximum allowable radiation exposure limits therefore clearly involves value judgments: unless the level is set at zero one is going to allow *some* extra cancer deaths; the determination of the worth of nuclear electricity in terms of human cancer deaths is clearly a value judgment.

Second, and more importantly, our immense lack of knowledge regarding the precise effects of radiation on living beings introduces an extra element of risk, and the evaluation of this risk and the setting of 'tolerable' risk levels are once again evaluative issues. The history of man's dealings with ionizing radiation inspires little confidence in the reassurances of some nuclear engineers that 'a little radiation will do no harm.' It is good to remind oneself of the fate of early radiologists who used to adjust their X-ray machines by holding their hand in the beam and noting the effect: the setting was considered right if the radiation failed to cause an appreciable reddening of the skin. Most of them died of bone cancer.[41]

Not only have the less discriminating advocates of nuclear power lost much credibility by the way they have misrepresented crucial issues, but an increasing amount of criticism is also addressed at the institutional structure in which the responsibility to make such morally controversial decisions is restricted to a small group. Critics of nuclear power programs often argue that such decisions belong in the public realm,[42] but they do not suggest any institutional alternatives to the existing ones, which are clearly inadequate. And merely to suggest opening up nuclear decision-making to public participation without at least indicating an institutional framework in which effective citizen participation can take place is rather less than satisfying. After all, it is the structure of the present government and semi-government institutions (in Canada the utilities, Atomic Energy of Canada Limited, the federal Atomic Energy Control Board, etc.) which has implicitly been so severely criticized.

Another conclusion brought up in the nuclear debate concerns the alleged novelty of the problems to which nuclear power gives rise. It is especially the existence of morally very controversial issues that is often singled out as lending a wholly new dimension to nuclear power planning. The issues most often stressed are the following: (1) the weapons potential of nuclear fuel and wastes; (2) the risk of accidents of enormous magnitude; (3) the damage caused by routine emissions from reactors, mines, fuel reprocessing plants, etc.; and (4) the fact that the radioactive poisons generated in the reactor tend to have very long lives and constitute a potential hazard for many generations to come. These characteristics, it is frequently argued, lend to nuclear power a new and unique quality: there are, in the words of one well-known anti-nuclear spokesman, 'technical bases for ethical concern.'[43]

But do these characteristics make nuclear power unique? And are technologies that give rise to ethical concern really all that new? For one thing, the military application of nuclear energy, which of course has a much longer history than civilian nuclear power, could have given rise to exactly the same objections.[44] The risks of proliferation of nuclear

weapons and of their illicit use can only be minimized by maintaining
strict secrecy, a practice admittedly more accepted in military spheres
than in the context of civilian power generation. The threat of 'accidents' of enormous magnitude is of course also inherent in the military
use of nuclear energy, and routine emissions due to the manufacture and
testing of nuclear weapons have up to now far exceeded those resulting
from civilian nuclear power programs. Indeed, the safety record of
military nuclear technology, as far as anything is known about it, is far
more abominable than that of the nuclear power industry. And, finally,
the threat posed by plutonium and other long-lived radioactive substances has been with us a great deal longer than civilian nuclear power.
The plutonium in the many thousands of nuclear bombs spread over
most of the world is certainly sufficient to wipe out life on earth many
times over. The threat that all this plutonium presents to posterity (or
even to there being any posterity) is rarely referred to in the nuclear
debate.

In short, it is hard to accept the claim that civilian nuclear power gives
rise to a radically new set of ethical problems, and one fails to see how
practices that have been condoned for years when the justification was
the gaining or maintaining of military supremacy are suddenly con-
demnable when the object is to generate electricity.

Neither can it be denied, in a much wider context, that technology will
always reflect the assumptions and preferences of the society or societal
group that developed it. In a society in which the development and
implementation of technological tools is the prerogative of a small
technological and industrial elite, it is quite conceivable that conflicts
over the desirability of technological projects will arise. And, seen in this
wider context of the politics of technological change, the problems
which are brought forward in the nuclear debate are not nearly as unique
as they are often made out to be. Indeed, to stress their uniqueness
detracts from perceiving the political nature of *all* technological planning and development; merely to point at the morally controversial
nature of the decisions involved does not solve anything.

There is one aspect, however, in which nuclear power does stand out,
and this, unfortunately, is the one point that is least often emphasized.
Centralized nuclear power generation calls not only for centralized
supervisory institutions charged locally with the prevention of accidents
and sabotage, and, at the international level, with deciding which
nations are trustworthy enough to be sold nuclear technology; it also
calls for maintaining such centralized control institutions until far

into the future. Reactors have a productive life of about twenty-five years, of which they have to spend a good fraction generating electricity in order to justify their construction costs. Moreover, they leave (long after they have been decommissioned) their radioactive remains and tons of exceedingly toxic wastes to be guarded for centuries, and this is not a task that can be organized on a neighbourhood basis.[45] In other words, massive nuclear development will make the centralized political system which created it virtually indispensable. This, it would seem, is not only unethical, it is a politically foolish course of action in view of the considerable problems created by the structure of our present institutions.[46]

THE POLITICS OF TECHNOLOGICAL CHANGE

The nuclear debate clearly suffers from a preoccupation with technical issues: the disputes about radiation hazards, accident risks, or the possibility of nuclear sabotage may lead to stricter regulations and safety measures, but they are unlikely to result in any fundamental changes in the way our society chooses its technological tools. To be sure, the 'wider societal consequences' of large technological projects are now frequently assessed, but the very fact that a distinction is made between what is technologically relevant and any 'societal' or 'human' implications is misleading and indicative of the mental outlook I am trying to expose. Technology is a societal tool and hence *all* its consequences are relevant to society.

Even the recognition that some of the problems regarding nuclear power generation have ethical dimensions has hardly been pursued. Instead of facing the political problem of resolving somehow the resulting conflicts, environmentalists tend to adopt a missionary attitude, assuming that moral enlightenment will suffice to convert the proponents of nuclear power programs. Furthermore, the emphasis that is often placed on the novel aspects and hazards of nuclear power generation has detracted attention from the much wider relevance of problems which are now seen as unique to nuclear power.

It is not difficult to see, however, that the three key issues behind the nuclear controversy (even if they are not often explicitly stated) are very general and recur over and again in our society, in connection with nuclear power, private transportation, agroindustry, medical technology, or any other a of a host of socially influential technologies.

The first of these issues concerns the steering of a society's

technological efforts: how society directs the development and application of technology. The sophisticated products of modern technology
have all required a considerable investment of scarce resources for their
development. In general, therefore, one may expect that the direction of
technological change will be strongly influenced by the perceived interests of the investors. Moreover, whereas the scientific basis for any
technological development may be far removed from everyday reality,
the design and development of technological products requires a large
number of decisions, most of which are made on full human evaluative
grounds. Thus, current ideological prejudices enter inevitably, and the
finished product will carry with it a set of implicit assumptions about the
societal environment in which it is supposed to function.[47]

The way in which technological development may almost impercepti-
bly change in direction is well illustrated by the tendencies towards
automation and centralization of production in our society. Not very
long ago the main function of technological tools was to lighten certain
jobs, thereby often creating new possibilities for the craftsman. Now,
however, there is an increasing tendency to develop technological tools
that completely eliminate jobs. Tasks which used to require skill and a
good deal of judgment are reduced to some standardized form suitable
for automation and mass production, and the role of the human operator
is reduced to that of serf to the machine. The machine's bland products
are guaranteed neither to please nor to displease anyone.

Technology is thus gradually being removed from the shop floor and
from people's everyday existence. Despite the variety of technological
products on which we rely, our insight into their operating principles is
minimal. In fact, commercial propaganda carefully substitutes pseudo-
scientific explanations for the real ones, and we are made to believe that
soap works because of its 'whitening power,' fertilizers give 'growing
power' to the plants, diseases are caused by 'germs' which can be killed
with 'antibiotics,' etc. In no previous culture was the average citizen so
ignorant about so much of his natural and man-made environment, and
indeed about his own body and mind, as in ours. For the provision, and
even the definition, of our everyday needs for keeping our tools, bodies,
and minds functioning correctly, and even for telling us what correct
functioning consists of, we have become almost completely dependent
on a small scientific and technological elite. And this elite, due to its own
fragmentation and specialization, is itself precluded from attaining insight in anything but a small field of specialist concern.

Automation and centralized production undoubtedly have the potential to free people from menial chores in order that they can spend their time and energy in more creative pursuits, and many products of technology constitute a genuine and valuable contribution to this end. Increasingly often, however, automation achieves just the opposite, by substituting standardized mechanical tasks for what used to be creative pursuits. Examples come easily to mind: centralized industrial production processes, including, for example, the mass-preparation of food, mass-produced 'entertainment,' machine 'teaching' in the electronic classroom, to mention only a few.

The decision to develop this kind of technology, rather than one more supportive of the need for engaging in meaningful and creative activities, was of course never consciously made, and to understand why technological innovation has taken this direction we have to inquire into the cultural and political circumstances in which it is pursued. This essay is not the place to explore this difficult and involved topic.[48] To say simply that the motivation was political or that the capital investments required to install machines were ultimately cheaper than human labour, would be an oversimplification; yet such motives clearly did play an important role. Neither can it simply be maintained that an explicit desire to control and manipulate consumers led to this development, although again such considerations do play an important role in corporate tactics. The fact remains, however, that the development of new technology is, in our society, the responsibility of a small group – a group whose interests are rarely representative of those of the population as a whole. The resulting conflicts, even if they are often ignored or repressed, constitute an important feature of our society.

A second important issue that arises in virtually any application of large-scale technological processes is that of establishing levels of allowable risk and damage. This has proved to be a particularly controversial issue in the context of nuclear power, but the problem is a very general one. Even if regulatory powers reside under some government agency, as is usually the case, the stringency of the regulations in actual practice depends strongly on what companies consider to be technically and economically feasible. In other words, regulations are rarely drawn on anything like independent criteria and an explicit trade-off of all costs against all benefits is rarely made. In addition, regulatory acitivities are often surrounded with an aura of scientific objectivity, which hides the extent to which value judgments and morally controversial decisions enter the process, and which simultaneously justifies its appropriation by a small group. Here too society denies the existence of conflicts rather than attempting to solve them.

The third and last key issue that arises in the nuclear controversy as well as in many other technological ventures concerns the distribution throughout society of the costs and benefits flowing from the projects. The economic market provides one mechanism for distributing some of the costs and benefits, but the frequent occurrence of externalities, especially where costs are concerned, leaves many cases in which the market is utterly unsatisfactory as an allocative mechanism. And the problem is anyway one of an entirely different order, for the adoption of the market as a mechanism for the distribution of scarce societal resources is clearly a morally controversial move itself.

The problems that surround centralized energy generation are symptomatic of the inability of the market mechanism to legitimize the distribution of costs and benefits of technological projects. Whereas the benefits of such projects are presumably spread widely and rather evenly, the costs, in the form of community disruption, expropriation of land, local pollution and health hazards, and so on, tend to be borne by the local population. When other nations and subsequent generations are concerned, as in trans-frontier pollution and plutonium hazards, the market mechanism loses all applicability.

Thus the market breaks down in just those cases in which its help would be most needed. In the allocation of costs the existence of externalities usually precludes a solution solely in market terms. Raw materials, an important category of commodities, are usually policy-priced rather than market-priced. In addition, the adoption of the market system itself and the existing distribution of buying power in the market remain to be justified.

Neither does our society possess any other mechanisms specifically designed to resolve the conflicts that arise over the distribution of societal costs and benefits. Our legal systems work only nationally and can only deal effectively with the grievances of individuals against other individuals. Authoritarian force is considered morally repulsive (and does not really provide a solution to any conflict except in so far as it removes a source). The existence of conflicts can be ignored, as is indeed often done, and this is becoming increasingly conspicuous as (potential) conflicts become more numerous.

Any serious policy alternative with regard to the guidance of technological change must therefore address the problem of how to provide an institutional framework for societal conflict resolution. This is no easy task and one certainly not to be attempted in this essay.[49] Suffice it here to point at one form of institutional organization which at least takes into account the inevitability of societal conflicts and the need to resolve these in a way compatible with the democratic ideal.[50] This alternative relies on a decentralization of societal decision-making and places itself thereby diametrically opposed to the existing centralized structures. A second characteristic of the alternative scheme is that it is dynamic at two levels. First, the planning objectives themselves evolve under the influence of mutual consultation. The notion of 'optimal decision' thus loses its meaning and has to be replaced with 'optimal trajectory.' Second, it is quite possible that permanent changes occur in the way goals are evaluated: societal decision-making is at best also a societal learning process, both through the information transfer among groups and through an increased understanding of external constraints. The theoretical basis of such systems of decision-making is just beginning to be explored and again the interested reader must be referred elsewhere for details.[51]

The ultimate motivation for exploring such alternative decision-making schemes is twofold. First of all one wants to avoid both the Scylla of decentralized incoherence and the Charybdis of coherence through central authoritarian control. It is my contention that only decentralized institutional structures that allow for iterative forms of societal decision-making along the lines suggested above will succeed in providing maximal local choice subject to the constraints of maximal global coherence and temporal continuity. And second, such an alternative may constitute an exploration of new ways of defining and subsequently satisfying societal needs. The ideal here is a progressive reconciliation between productive and consumptive roles in society – not by anonymous market signals, but by a societal process of joint decision-making and learning.[52]

The implications of both these abstract goals for technological innovation should be clear. Alternative institutional structures such as the one sketched above are not wholly utopian, as is evident from some recent Yugoslavian and Chinese experiments.[53]

CONCLUSION AND SUMMARY

The ethical problems and conflicts of interest that are at the base of the nuclear power controversy are becoming so conspicuous that we are forcefully reminded of the political aspects of technological change: how does our society choose which technology to adopt or develop in order to satisfy some societal need, who determines what our needs are, and how is the course of action taken subsequently justified?

To trace this development in some detail we have discussed the standard justification for nuclear power programs, viz. the prediction of future demand for more electricity (load forecasting), combined with the conclusion that, on economic and technical criteria, nuclear power is preferable over other forms of electricity generation. We have found both justifications severely wanting, and concluded that future demand for electricity as well as the future economic or technical superiority of any one of several potential energy sources is strongly dependent on present and future policy. It is also clear that in the past military and corporate interests and an ideologically based predilection for centralized energy technology were important factors in the decision to adopt nuclear energy. These latter factors rarely receive the attention they deserve.

We subsequently examined the arguments commonly adduced against the adoption of nuclear power generation and concluded that: (1) although the uniqueness of the nuclear hazards is often emphasized, the general problems created by nuclear power generation are certainly not new, but are typical of the centralized mode of commodity production which now characterizes market society; and (2) that virtually all antinuclear arguments rest on ethical (not technical) objections. The nuclear controversy is therefore in essence a political problem, that of who decides on morally controversial issues and how. It is also clear that these are important issues, for what is at stake is the very form of our social order as well as that of many future generations.

The unique level at which these issues pose themselves, however, should not blind us to the generality of the problem. There are, we have suggested, three key issues in the nuclear controversy: (1) how is the development and adoption of new technology to be guided – by a small elite (and therefore consciously or not in its interests) or according to some form of general consensus? (2) how does society assess what are tolerable risks – whether for small groups or for society as a whole? (3) how are costs and benefits defined and how are they distributed over society?

These problems arise over and over in our society, in connection with nuclear power, private transportation, agroindustry, medical technology, or any of a host of other societally sensitive technologies. But, as the nuclear example so clearly shows, the existence of conflicts or potential conflicts is often denied, sometimes in extremely subtle ways (scientistic arguments, refusal even to consider certain issues, etc.). And indeed market society possesses no mechanism for societal conflict resolution save brute authoritarian force and the *dictature* of the market. The former is regarded as morally repulsive and cannot openly be used; the latter cannot effectively deal with many key problems (the existence of externalities) and is based on highly controversial ethical assumptions.

Any serious policy alternative with regard to the guidance of technological change must therefore address the problem of providing an institutional framework for societal conflict resolution. We have only hinted at the institutional reforms that this would require, and we briefly described one alternative that involves decentralized iterative forms of decision-making. Such an alternative, however, will be of little more than academic interest once the widespread adoption of nuclear power has made centralization and a high technological dependence mandatory.

NOTES

1 Government of Canada, *An Energy Policy for Canada*, 2 vols. (Ottawa 1973); see especially I, section II, chap. 5

2 See, for example, J. Gofman and A. Tamplin, *Poisoned Power* (Emmaus, Pa. 1971); R. Gilette, 'Nuclear Safety' (four-part series), *Science*, 4051–4 (Sept. 1972); D.F. Ford *et al.* (Union of Concerned Scientists), *The Nuclear Fuel Cycle* (San Francisco 1974)

3 For example, S. Novick, *The Careless Atom* (Boston 1969)

4 See, for example, A.B. Lovins and J.H. Price, *Non-nuclear Futures: The Case for an Ethical Energy Strategy* (Cambridge, Mass. 1975)

5 D.D. Comey, 'The Legacy of Uranium Tailings,' *Bulletin of Atomic Scientists* (Sept. 1975), 43–5, calculates that the total death toll directly due to radon emanation from US mine tailings during the rest of this century will be of the order of twelve million, albeit spread over the next 80,000 years.

6 Instructive in this regard are the documents submitted by Ontario Hydro to the Royal Commission on Electric Power Planning. See especially Ontario Hydro, *Long-Range Planning of the Electric Power System*, Report no 556SP (Toronto, 1 Feb. 1974); Ontario Hydro, *Public Participation*, memorandum to the RCEPP (Toronto, March 1976). Despite the rosy picture painted in these documents ('... Hydro has involved the public in varying degrees in all projects undertaken since 1972') the possibilities for effective public input are really minimal and concern chiefly matters of detail. One can get a glimpse of an explanation for this state of affairs if one reads the document on public participation cited above. It ends as follows: 'Often, public participation has been against the result of the study (a site or line location) rather than the study process. This suggests that a more realistic understanding of the role of public participation is required *so that the process of involving the public is not considered a replacement for decision making*' (my emphasis). As is pointed out elsewhere in the report, 'Public

participation means different things to different people'! One can indeed observe a tendency towards even greater secrecy in such important matters as nuclear safety and capital borrowing.

7 See, for example, Lovins and Price, *Non-nuclear Futures*
8 For example, W.C. Patterson, *Nuclear Power* (Harmondsworth 1976)
9 R. Gardner, 'Working Class Consciousness and the Structure of Power in Canada,' in A. Turrittin, ed., *Proceedings of the Workshop conference on Blue Collar Workers and Their Communities* (Toronto 1976), 182–208, and several other papers presented at the same conference.
10 See *An Energy Policy for Canada*
11 Ontario Hydro, *Load Forecasting*, memorandum to the RCEPP (Toronto, May 1976)
12 See, for example, Ontario Energy Board, *Ontario Bulk Power Rates for 1976, Part II* (Toronto, Feb. 1976). It should be noted here that the most recent demand projections of Ontario Hydro are significantly lower than previous ones. The fact remains, however, that the utility continues to project demand, rather than trying to implement a well-defined energy policy. Such a policy is, of course, still sadly lacking.
13 Some passages in the final report of the Select Committee of the Legisture Investigating Ontario Hydro, *A New Public Policy Direction for Ontario Hydro* (Toronto, June 1976), point in this direction. For more extensive remarks on these issues, see C.A. Hooker and R. van Hulst, 'A Preliminary Study of the Conceptual and Institutional Structure of Energy Policy Making in Ontario and Its Policy Alternatives' – available on request from the RCEPP, and P.L. Cook, 'Energy Policy Formulation,' in W. Leontief, ed., *Structure, System and Economic Policy* (Cambridge 1977)
14 W. Leiss, *The Limits to Satisfaction* (Toronto 1976)
15 By 'market society' I understand a society in which the economic market either constitutes the major societal mechanism for the allocation of scarce resources or at least figures importantly in the legitimation of the way in which resources are allocated. Thus defined, 'market society' does not coincide with the economical construct of the free and competitive market. The latter is a theoretical 'ideal,' the former a characterization of a form of societal organization. For the development of market society, see K. Polanyi, *The Great Transformation* (Boston 1957)
16 By liberal economists, at least. Marxist economics has, of course, drawn attention to this fact from its very beginnings.
17 We have attempted to do so in C.A. Hooker and R. van Hulst, 'System and Value: The Systematic Structure of Socio-Environmental Problems' (to be published). See also our joint paper in this volume.
18 Roland Barthes, *Mythologies* (New York 1972), characterizes myths as 'transforming history into nature,' that is, as suggesting a natural necessity for processes that result from historical actions.
19 For a detailed critique of the (conventional) neo-classical approach to economics, see M. Linder, *Anti Samuelson*, 2 vols. (New York 1976). Another excellent critique is M. Hollis and E.J. Nell, *Rational Economic Man* (London 1975). Also relevant is H. Albert, *Marktsoziologie und Entscheidungslogik: Ökonomische Probleme in soziologischer Perspektive* (Berlin 1967)
20 See for example, Ontario Hydro, *Generation-Technical*, submission to the RCEPP (Toronto, March 1976)
21 Similarly, a society relying on rail transport (trains, streetcars) and bicycles for personal transportation (such as most western European countries around the beginning of the century) would be hesitant to invest in the construction of highways and car

factories, and the demand for private motorcars would be minimal, initially. (Historically the increased demand for cars in western Europe coincided with a rapid decline in the quality of public transportation.)

22 See, for example, Canada, Energy, Mines and Resources, *An Energy Strategy for Canada: Policies for Self-Reliance* (Ottawa 1976)

23 A good introduction is A.K. Dasgupta and D.W. Pearce, *Cost-Benefit Analysis: Theory and Practice* (London 1972).

24 C. Nash, D.W. Pearce, and J. Stanley, 'An Evaluation of Cost-Benefit Analysis Criteria,' *Scottish Journal of Political Economy*, 22 (June 1975), 121–4

25 On cargo cults, see, for example, Mary Douglas, *Natural Symbols: Explorations in Cosmology* (New York 1973)

26 A. Cockburn and J. Ridgeway, 'The Sunny Side of the Street,' Review of Commoner's *Poverty of Power, New York Review of Books*, 23 (5 Aug. 1976), 44

27 See J. Habermas, 'Technology and Science as Ideology,' in his *Toward a Rational Society*, trans. J.J. Shapiro (Boston 1970)

28 See, for example, J. Ridgeway, *The Last Play* (New York 1973)

29 See, for example, L. Scheinman, 'The Nuclear Safeguards Problem,' in Ford, *Nuclear Fuel Cycle*, 57–74; Stockholm International Peace Research Institute (SIPRI), *Nuclear Proliferation Problems* (Cambridge, Mass. 1974); L.D. DeNike, 'Radioactive Malevolence,' *Bulletin of Atomic Scientists*, 30, 2 (1974), 16–20

30 National Academy of Sciences (US), *The Effects on Populations of Exposure to Low Levels of Ionizing Radiation*, BEIR report (Washington, DC 1972); UN Scientific Committee on the Effects of Atomic Radiation (UNSCEAR), *Ionizing Radiation: Levels and Effects*, 2 vols. (New York 1972)

31 See Canada, Energy, Mines and Resources, *1975 Assessment of Canada's Uranium Supply and Demand* (Ottawa, June 1976), for recent estimates. Although domestic requirements until well into the twenty-first century can be met from already delineated resources two factors give rise to concern: first, the market price of uranium has increased tenfold in the last decade and at present prices it will certainly be attractive for Canadian industry to export as much uranium as the government will allow; and, second, the report referred to shows that doubling of the price of uranium increases Canada's measured and indicated uranium reserves only slightly.

32 See the September 1975 issue of the *Bulletin of Atomic Scientists*, devoted to this problem. For a technical critique of the commonly used arguments purporting to show that accident risks are negligible, see T.P. Speed, 'Negligible Probabilities and Nuclear Reactor Safety: Another Misuse of Probability?' to be published.

33 This is the 'Faustian bargain' argument. See A. Weinberg, 'Social Institutions and Nuclear energy,' *Science*, 177 (7 July 1972), 27–34

34 See, for example, Lovins and Price, *Non-nuclear Futures*

35 For example, E. Sternglass, *Low Level Radiation* (New York 1972). See also the review of this book by A.R. Hoffman and D.R. Inglis, 'Radiation and Infants,' *Bulletin of Atomic Scientists* (Dec. 1972), 45–52. It should also be noted that Canadian radiation standards are far less stringent than American ones.

36 Cockburn and Ridgeway, 'The Sunny Side of the Street'

37 For example, in the glossy brochures in which Atomic Energy of Canada Ltd. praises its brainchild, the CANDU reactor.

38 See, for example, *Ramparts'* analysis (Aug. 1974, pp. 26–7) of a typical utility advertisement.

39 Another favourite comparison is with radiation received from medical or dental X-rays.

It is now quite generally recognized, however, that the latter are often indefensibly high. See K.Z. Morgan, 'Never Do Harm,' *Environment*, 13, 1 (Jan.–Feb. 1971), 28–38

40 See, for example, the BEIR and UNSCEAR reports

41 K.Z. Morgan, 'History of Damage and Protection from Ionizing Radiation,' in Morgan and J.E. Turner, eds., *Principles of Radiation Protection* (New York 1967)

42 For example, Lovins and Price, *Non-nuclear Futures*

43 *Ibid.*

44 And, to some extent, it has, as a glance at past volumes of the *Bulletin of Atomic Scientists* shows. Environmentalists, nevertheless, tend to ignore the hazards posed by the military use of nuclear power. It is rarely realized, for example, that the US military program has already created large quantities of highly dangerous nuclear wastes: in fact, the reactor wastes from the US civilian nuclear power program will not reach 10 per cent of this total before the mid-1990s. See D. Leslie, 'Redressing the Nuclear Balance,' *New Scientist*, 71 (30 Sept. 1976), 700–1

45 G. Bailie, 'Nuclear Social Order 239,' *Liberation*, 19 (July–Aug. 1976), 18

46 See Hooker and van Hulst, 'System and Value'

47 See, for example, D. Dickson, *Alternative Technology and the Politics of Technical Change* (Glasgow 1974)

48 See, however, Habermas, 'Technology and Science as Ideology'; L. Tribe, 'Technology Assessment and the Fourth Discontinuity,' *Southern California Law Review*, 46 (1973), 617–60; and Dickson, *ibid.*

49 See Hooker and van Hulst, 'Preliminary Study' and 'System and Value'

50 The following section owes much to B. Boucon, J. Bourles, J.-H. Lorenzi, and B. Rosier, *Modèles de planification décentralisée* (Grenoble 1973). See also T. Baumgartner, T.R. Burns, P. DeVille, and L.D. Meeker, 'A Systems Model of Conflict and Change in Planning Systems with Multi-level, Multiple Objective Evaluation and Decision Making,' *General Systems*, 20 (1975), 167–83

51 In particular, see Boucon, *ibid.*; A. Rapoport, ed., *Game Theory as a Theory of Conflict Resolution* (Dordrecht, Holland 1974); M.D. Mesarovic, D. Macko, and Y. Takahara, *Theory of Hierarchical, Multi-Level Systems* (New York 1970); T. Burns and W. Buckley, eds., *Power and Control: Social Structures and Their Transformation* (London 1975)

52 Boucon, *ibid.*; see also Leiss, *Limits to Satisfaction*

53 See, for example, G. Hunnius, 'The Yugoslav System of Decentralization and Self-management,' in C.G. Benello and D. Roussopoulos, eds., *The Case for Participatory Democracy* (New York 1971); R. Berger, 'Planification à la Chinoise,' *Le Monde diplomatique* (March 1973)

9
Energy, ecology, and politics

Robert Macdonald

Industrialization is a precariously grafted scion upon man's age long existence. The outcome of this experiment is still hanging in the balance.

Karl Polanyi, 'Our Obsolete Market Mentality'

THE ENVIRONMENTAL CRISIS

Within the last decade our society has been encountering a growing number of challenging public policy issues – issues that have manifested themselves as a series of 'crises,' including a pollution crisis, a food and population crisis, water and mineral crises, a crisis in our cities, a crisis in the availability of agricultural land, an on-going series of economic crises, and an energy crisis. What may well distinguish this set of crises from previous public policy issues is the speed with which they have come upon us, their scale and complexity, and their apparent intractability. The evidence is growing that existing institutions, both public and private, are by and large unable to address these issues effectively. After many years of effort they remain largely unresolved.

Although often perceived as being unrelated, and still largely treated as such by our institutions, there is an extensive body of data and analysis that argues that these crises are in fact intimately related, with a common origin lying in the essential nature of our modern industrialized society and its relationship with the environment. Exponents of this point of view argue that, taken together, these issues constitute an evnironmental[1] crisis of such severity as to constrain severely our present patterns of social and economic activity and to threaten the future well-being of our society.[2]

The literature that examines the nature and origins of this environmental crisis is marked by considerable controversy. One controversial analysis was that undertaken for the Club of Rome, an international group of scientists, industrialists, and civil servants whose computer-based study of the environmental crisis was published in popular form as *The Limits to Growth*.[3] This project attempted to determine future implications for society of the interactions between resource limitations, population growth, pollution, food and arable land shortages, and industrial capital requirements, and concluded that imminent constraints exist to present patterns of economic and social activity characterized by exponential patterns of growth in material consumption, pollution, and population.

Criticisms of *The Limits to Growth* conclusions have been both methodological and substantive.[4] However, more sophisticated later analyses conducted by the Club of Rome have reached essentially similar conclusions with quite different analytical methods, while support for the report's substantive conclusions have come from many independent sources.[5] In general terms, these analyses conclude that industrial civilization is facing an interrelated set of limits and constraints on its activities, consisting not only of physical limitations such as a 'Malthusian scarcity of food ... impending shortages of mineral and energy resources, biospheric or ecosystemic limitations on human activity' but also of social and institutional constraints including 'limits to the human capacity to use technology to expand resource supplies ahead of exponentially increasing demands (or to bear the costs of doing so).'[6]

While the controversy continues, these studies appear to pose a challenge, which must be taken seriously, to the nature of current patterns and future trends of social and economic activity associated with the industrial model of economic growth and development. The conclusion emerging from these analyses is that society must, in a relatively short period of time, both recognize and in some appropriate manner respond to the environmental crisis, if serious disruption of an ecological, economic, social, and political nature is to be avoided. The study of the environmental crisis is truly an interdisciplinary task. Ethics, religion, and all branches of the social and physical sciences must be brought to bear in addressing these issues. Only through the most creative application of knowledge, both new and rediscovered, can mankind hope to succeed in establishing a desirable and sustainable man-environment relationship.

In this regard, the discipline of ecology has been of central importance in expanding our understanding of the nature of the complex systems of the biosphere which embrace and support all life on this planet, the potential that man's activities have for disrupting these systems, and, in turn, the consequences for man of their disruption.[7] Ecology establishes guidelines within which a society that hopes to achieve a sustainable relationship with the biosphere must circumscribe its activities.[8] In this sense, ecology has been termed the 'critical science': 'The critical edge of ecology derives from its subject matter – from its very domain. The issues with which ecology deals are imperishable in the sense that they cannot be ignored without bringing into question the survival of man and the survival of the planet itself.'[9]

The following discussion examines but one component of the environmental crisis – the energy crisis – from the point of view of the critical science, ecology. Policy concerned with energy is but one of the many public policy issues which must be examined in the light of the environmental crisis; but the essential and ubiquitous role of energy in a modern, industrialized, and technologically sophisticated economy results in a situation in which an examination of the nature and the origin of problems in the energy sector serves to illuminate the nature of a wide variety of public policy issues within the environmental crisis.

ENERGY AND THE ENVIRONMENTAL CRISIS

An analysis of the energy crisis immediately reveals that the problem facing society is not one of running out of energy. On the contrary, at least on a planetary scale, there is no absolute scarcity of potential energy resources. Immense flows of solar energy are intercepted every day by the earth; very large amounts of energy exist in the form of coal deposits, tar sands, heavy oils, and oil shales; even the granite rocks of the earth's crust contain very large amounts of energy in the form of uranium, while the oceans of our planet constitute a virtually inexhaustible source of fuel in the form of the nuclear fuel deuterium.[10] Rather, the crisis stems from the growing costs of an ecological and social nature associated with attempting to meet society's already large and exponentially growing demand for energy.

Although Canada is not running out of energy in an absolute sense, what is occurring in the oil sector (our single most important fuel, constituting nearly one-half of total Canadian energy requirements) is that the production capacity of the readily available and therefore relatively inexpensive oil reserves of western Canada is expected to decline slowly but steadily, starting in the early 1980s.[11] In 1977, Canada required net oil imports of about 400,000 barrels of oil a day. Even assuming a slowly growing economy and the success of present (modest) conservation efforts, oil import requirements are likely to grow to at least 600,000 barrels of oil per day by 1990. This requirement for imported oil not only leads to a serious drain on Canada's economy but potentially places Canada in a politically vulnerable position, as most of the imported oil will be coming from the Middle East. Moreover, Canada's demand for international oil will be increasing at a time when it is believed that, on a global basis, demand for oil will exceed the supply.[12] Such a situation is fraught with uncertainty in terms of future costs and availability.

Even this limit of about 600,000 barrels per day on oil imports will be achieved only by means of a massive program of capital investment in energy production facilities. Between 1976 and 1990, according to federal government policy documents, Canada will have to spend more than $180 billion to provide additional domestic energy supplies.[13] About one-half of this will go to petroleum and natural gas exploration and development, and to coal. The remainder will be required for the development of electrical power. An examination of these existing and proposed energy supply projects throughout Canada illustrates the impacts of such projects both upon the biosphere and upon society. In general, these impacts stem from the scale of these energy supply projects, and their location and/or the nature of the technology involved. Proposed energy schemes include a program of development of oil and gas resources in the frontier regions of Canada: the Arctic, particularly offshore in the Beaufort Sea, the Arctic islands, and the Mackenzie River Delta, and offshore on the Atlantic east coast. Concern has been expressed by ecologists that a large oil 'blow out' at an offshore drilling site – the accidental, uncontrolled escape of large amounts of oil (and, to a lesser extent, natural gas) under high pressure – might continue unchecked for months and even years before a nearby relief well could be drilled, due to the difficulties of constructing and operating drilling equipment in the severe climate of the North.[14] Such an accident could release very large amounts of oil into the sea where it would disperse under the ice and be effectively uncontainable; this would have a very severe impact upon the rich mammal, bird, and fish ecosystems of the region. Moreover, the extremely slow decomposition of oil at low temperatures would result in this pollution and its impacts persisting for long periods of time, perhaps decades. Climatologists have suggested that the dark petroleum liquid might be trapped within the relatively thin Arctic Ocean pack ice, subsequently migrating to the surface of this ice layer and there, through a reduction of the albedo of this ice cover, trigger a process leading to its irreversible and rapid melting.[15] Major, unpredictable, and uncontrollable global climate changes are believed likely to result from the loss of the Arctic ice cover.

A major requirement for the development of Arctic oil and gas resources is a means for transporting them southwards. Two major pipeline corridors have been proposed: in the western Arctic, from the Mackenzie Delta southwards along the Mackenzie River Valley into Alberta; and in the eastern Arctic, stretching from the Arctic islands southwards along the west side of Hudson's Bay into Ontario.[16] Northern ecology is fragile, and the potential for ecological disruption by northern pipeline construction, as well as the creation of severe impacts upon the economies and culture of native peoples of these regions, has led to considerable concern among ecologists and native peoples' organizations alike. Beginning in 1975, a commission of inquiry headed by Thomas Berger undertook a comprehensive analysis of the costs and benefits for the region of the construction of a natural gas pipeline along the Mackenzie River Valley, linking Alaskan natural gas and Mackenzie River Delta natural gas resources to southern markets. In the face of intense pressure from all levels of government and multinational energy corporations, this inquiry concluded that, at this time, the costs of the pipeline project both to the native peoples and the ecology of the region outweighed its purported benefits. The commission recommended to the federal government that the construction of the pipeline be deferred for at least ten years and that certain ecologically vulnerable areas of this northern region be closed indefinitely to pipeline construction.[17]

Energy resource development projects such as western mining of coal and the Athabasca tar sands projects involve large-scale strip mining activities and the production of large amount of wastes. Such activities, if conducted on the large scale apparently necessary to meet our projected future energy requirements, could lead to severe environmental disruption. One of the largest energy projects in the world, both in terms of capital requirements and in terms of the land area affected, is Quebec's James Bay hydroelectric power development. Affecting thirty thousand square miles of northern river watersheds and projected now to cost $25 billion to complete, large ecological costs appear inevitable. In marked contrast to the exhaustive impact analysis of the Berger inquiry for the Mackenzie Valley pipeline, the James Bay development is notable for the virtual non-existence of any serious impact assessment.[18]

Coal, petroleum, and natural gas are all hydrocarbons. An inevitable consequence of the combustion of hydrocarbon fuels is the production of carbon dioxide, a gas that contributes significantly to the heat trapping or 'greenhouse' effect of the earth's atmosphere. In addition, all energy utilized by society, regardless of its initial form, ends up being released as heat to the environment. Scientists have presented evidence suggesting that within fifty to seventy-five years, if established patterns of growth in global energy consumption continue, the addition to the atmosphere of carbon dioxide from the burning of fossil fuels, the introduction of particulate matter into the atmosphere from energy fuelled industrial and transportation activities, or the addition of heat to the environment as the inevitable side-effect of the use of energy, or some yet unknown, possibly synergistic, interaction among these three, may well produce unpredictable, uncontrollable, and irreversible changes in the climate on a global scale.[19]

This situation presents world society potentially with an extremely serious problem. Man's agricultural activities, the fundamental basis for the existence of society, are precariously dependent upon the continuation of *existing* climate conditions. Even small changes in global patterns of weather could produce a disastrous famine. Canada, for example, perceived as one of the bread baskets of the world, is only marginally an agricultural nation because of the severity of its climate. A small temperature shift or a change (due to increased cloudiness) in the amount of sunlight reaching the ground could critically reduce the frost-free growing season in western Canada, jeopardizing its ability to grow wheat.[20]

While present models of global climate are inadequate to predict reliably the consequences of man's energy-related interventions into the mechanisms of the biosphere-controlling climate, the evidence suggests that in effect the industrialized nations of the world, as the chief consumers of energy, are conducting a global climate 'experiment' over which they have no control and little understanding of the consequences. Unfortunately, by the time that conclusive evidence of climate modification becomes available, it will be largely impossible to reverse the experiment because of the many decades of lead time that would be required to make a large-scale change in global patterns of energy use and the irreversibility of man's addition of wastes and heat to the atmosphere. Thus potential and – relative to the lead time required to alter man's patterns of energy use – imminent climate modifications may well present man with an upper limit on the use of fossil fuels; given the inevitable production of heat associated with nuclear fission (discussed in the following section) or fusion energy, such constraints may place an upper limit on all non-solar derived forms of energy.[21]

The disappointing results of frontier oil exploration, the heavy costs and lengthy construction time of tar sands facilities, and the costs and future uncertainties as to the availability of international oil, urgently argue the need to develop new sources of energy to replace oil. Over the long term, natural gas cannot serve as such a replacement, for the life expectancy of Canada's natural gas reserves is only a decade or so longer than those of oil (though natural gas does have a critical role to play in bridging energy supply short falls during the transition period to non-hydrocarbon energy sources); coal, a solid, is an inconvenient energy form and the technology to convert coal into more convenient liquids and gases seems unlikely to be implemented on a large scale in the near future.

At the present time, two major non-fossil fuel energy forms appear to hold the potential to serve as a replacement for oil: electricity (largely generated from nuclear energy) and a variety of solar derived, renewable energy resources.

THE NUCLEAR ELECTRIC OPTION

Fully one-half of the capital required by the energy sector in Canada to 1990 is expected to be devoted to the construction of electrical generating facilities. These include hydroelectric facilities such as Quebec's James Bay project and coal burning thermal stations. A substantial portion of this $90 billion will be for generating plants powered by nuclear fission energy.[22] The province of Ontario is in the vanguard of the development of this nuclear energy option. The importance attached to nuclear energy by the provincial Ministry of Energy is evident in this statement by R.M. Dillon, at that time deputy minister of energy (and now deputy provincial secretary in the Provincial Secretariat for Resources Development): 'In these days of uncertainty over the availability of fossil fuels and considering the magnitude of the decisions that must be taken, outside of Ontario, to assure reliable, long term supplies, it is not surprising to find Ontario looking to electricity to supply as much energy as possible and to develop nuclear power as far as technically and economically feasible ... The view is also widely held that electricity will take over as the prime form of energy, this electricity being produced initially by coal and uranium fuel stations ... Present forecasts by Ontario Hydro call for the installation of 20,000 megawatts of (nuclear) electrical generating capacity to be in service in 1990. This amounts to 40% of the total installed capacity and certainly indicates a strong move towards a nuclear based electricity system. The feeling in Ontario is that we have already taken a first step towards a nuclear electric society.'[23]

The development of nuclear energy in Ontario is being undertaken by the provincial utility, Ontario Hydro, in co-operation with Atomic Energy of Canada Limited. Research and development on the commercial application of nuclear energy has been underway by these two organizations for more than twenty-five years, with the result that in 1974 Ontario Hydro had an installed commercial nuclear generating capacity of about 2000 MW, located at Pickering (MW = megawatts or one million watts, a measure of the power output of the generating station). In that year, and in greater detail in subsequent years, Ontario Hydro brought forward a set of proposals for dramatically expanding its electrical generating capacity which would involve the construction of the equivalent of twenty additional Pickering-scale plants by the year 2000, with the equivalent of more than twenty-three additional such plants under construction by 2008.[24]

Paralleling the concern expressed in other countries over nuclear electric programs, considerable controversy was generated by the Ontario Hydro proposal, stemming both from its capital costs as well as from the social and environmental costs which, critics argued, would inevitably accompany the program. In response to this controversy, a Royal Commission on Electric Power Planning was constituted in 1975 to examine Ontario Hydro's proposed expansion plans. This commission has served as a forum for advocates and critics of nuclear power (and electrical generation programs in general) alike to present their viewpoints.[25] Of particular interest here are the ecological implications associated with the nuclear component of the Ontario Hydro proposal.

Environmental criticism of nuclear power programs has focused on the toxic, long-lived, radioactive wastes associated with all parts of the nuclear fuel cycle, including the mining of uranium, the processing of the uranium into fuel, the operation of the reactor, and the disposal of the wastes produced in the nuclear fission process. On one hand, the development of techniques for ensuring the separation from the biosphere, for the requisite period of time, of high level radioactive materials produced in nuclear reactors (which in some instances can remain harmful for periods of 100,000 years or more) presents particular difficulties (it is no consolation that many wastes of our industrial society, including mercury and certain other chemicals, also remain toxic for very long periods of time). On the other, the effect on man of low levels of radiation of the sort experienced by workers in the normal operation of all parts of the nuclear fuel cycle remains unknown. Concern has been expressed as to the possibility of mishaps, either accidental or as a result of sabotage, leading to the serious, long-term radioactive contamination of large areas surrounding a nuclear power station. Critics of the social impacts of the proposed nuclear program noted the disproportionately large amount of provincial funds required to undertake the proposed expansion, the dangers of allowing an already large and powerful institution (Ontario Hydro) to become even larger and more powerful in terms of its economic and political influence, and the contribution of Canadian nuclear technology, being actively marketed abroad, to the proliferation of nuclear weapons.[26]

Critics further pointed out that the (by no means insignificant) problems and hazards associated with the existing, relatively small, nuclear commitment would escalate in proportion to the size of the proposed program. Moreover, it now appears that if nuclear energy is to play a role beyond the year 2000, uranium resource constraints require that the present, relatively straightforward, nuclear fuel cycle employed in the present generation of CANDU reactors be replaced by a yet to be developed, considerably more complex, nuclear technology, involving a complicated fuel cycle that requires the recycling of nuclear waste products and the production of plutonium (a toxic, nuclear fuel suitable for nuclear weapon construction), all of which increases significantly the potential for environmental pollution by radioactive materials.[27]

Nuclear power, as yet on a relatively small side, is a reality in Ontario. In 1978 a total of 3500 MW of nuclear-generated electrical power were produced, with a further 9000 MW under construction or committed. In the face of the criticisms being encountered by nuclear technology both in Canada and abroad, the urgent question facing Ontario is whether or not there might exist other more desirable energy supply options open to the province in its search for a replacement for oil.

THE RENEWABLE ENERGY OPTION

A second source of energy appears to offer the potential of serving as the basis for society's future energy requirements: the sun. Solar energy is available in a number of different forms: directly, in the form of heat and light; as mechanical energy in the motion of air masses or wind energy;

as chemical energy stored in biological materials or 'biomass,' including wood, plant, and animal wastes and the (organic) garbage of our cities. A form of solar energy presently harnessed on a large scale, by hydroelectric installations, is the energy of falling water.[28]

Many different types of technology already exist for harnessing the various forms of solar energy. In Ontario, flat plate solar energy collectors for space and water heating of buildings appear to be a particularly attractive alternative to the use of nuclear electricity for the same purposes.[29] Indeed, the skilful design of new buildings and the renovation of existing buildings offers the potential for meeting a large portion of building space heating requirements by simple (but by no means unsophisticated) 'passive' solar heating methods: the installation of large, southern-facing windows in a well-insulated building to allow the penetration of sunlight; floors, walls, or ceilings designed to store heat; and insulated window shutters to trap the stored heat at night. The conversion of a variety of biomass materials, including forest products, urban garbage and sewage, into the (liquid) fuel methanol offers the possibility for fuelling an energy efficient transportation sector in the post-petroleum era.[30] Even considerable hydroelectric potential remains in the form of small hydro sites throughout the province, which could contribute significantly to provincial electrical needs in an energy strategy confining the use of electricity to the relatively small number of applications that require the unique properties of this energy form: lighting, small motors, electronics, etc., but *not* low temperature heat applications such as space heating, water heating, and low temperature industrial applications.

Only within the last five years, in the face of the demise of the oil era and growing concern over the nuclear option, have governments in Canada and abroad begun to consider seriously the solar energy option. Indeed, significant – but still small relative to that committed to the nuclear option – federal funding for the development and the implementation of solar energy technologies has been forthcoming only as recently as mid-1978.[31] Nevertheless, in this relatively short time, primarily because of the very large solar program underway in the United States, a small but significant solar energy industry has begun to emerge in Canada. There is now reason to believe that, with an effective, large, government support program, a variety of solar energy technologies could provide, by the year 2000, amounts of energy equivalent to those projected to be made available (in the form of electricity) by the nuclear option.[32] Indeed, if combined with imaginative and extensive energy conservation measures, it appears that in the early decades of the next century (less than fifty years from now) our national solar energy resources could meet most of Canada's energy requirements.[33]

In terms of its ecological impacts, the solar energy option offers the possibility of an energy supply system with minimal ecological consequences. Solar energy is the energy form that powers the processes of the biosphere. Very large amounts of solar energy, far in excess of society's current demands for energy, are daily incident on the planet. Consequently, no serious imbalances are produced when man intercepts a small portion of that energy, harnesses it for his purposes, and then releases it, as heat, back into the biosphere. The use of solar energy for heating homes, for example, avoids the host of problems associated with the exploration for and production of Arctic oil or natural gas, its transportation southwards, and its subsequent combustion in our cities. Critical in terms of its significance for the controversy surrounding the nuclear program in Ontario is the fact that solar energy technologies completely avoid the intractable problems of radioactive waste and large-scale heat production associated with nuclear generating plants. No climate altering air pollution or thermal pollution are created by the direct harnessing of solar energy.

The ecological impacts of large-scale forest management for the production of methanol are, potentially, less benign. The record of government and industry forest management for lumber and pulp and paper production has been dismal. However, it does appear possible to conceive of a cyclical forest management process which would, on a sustainable basis, allow the growth of trees, their harvesting for conversion to methanol, and the return of the mineral residues of the conversion process to the forest. Wind energy, which may have considerable potential for generating electricity in regions of Canada with strong, steady wind regimes, has no known serious ecological consequences. Even small-scale hydroelectric installations tend to distribute their ecological impacts in comparison with very large hydroelectric projects such as Quebec's James Bay scheme.

In addition to the ecologically benign nature of the technologies for harnessing solar energy, the energy from the sun is truly renewable, on any human time-scale. A society based upon renewable energy could indeed count on having an assured, long-lived, energy resource base. It is important to recognize that solar energy could be harnessed by technologies which could have potentially serious environmental impacts. For example, in the United States it has been proposed that large areas of land could be covered with mirrors, which would focus sunlight onto a tall steam boiler tower, for the production of electricity.[34] Such a scheme very much continues in the centralized, capital-intensive model of a nuclear generating program. Another scheme involves the construction of very large, orbiting space stations, which would gather solar energy in space and beam it to earth in the form of microwaves. Huge forest harvesting operations for the production of methanol could lay waste our forest resources. Large wave energy generating facilities, off-shore in our maritime areas, could also have both direct and indirect ecological consequences.

Thus it is clear that the adoption of solar energy may well be a necessary but by no means sufficient condition for the development of an energy strategy consistent with a sustainable man-environment relationship.

ENERGY, ECOLOGY, AND POLITICS

Arguably, the major challenge facing global society today, in the face of the environmental crisis, is that of determining the range of possible answers to the questions: 'what is the nature of a desirable society living in harmony with the biosphere?' and 'what measures can be taken today which are appropriate for moving towards such a society?'[35] Terms such as 'desirable society' and 'appropriate measures' are both vague and, inevitably, highly subjective. Immediately on considering these, a host of further questions arise: 'desirable for whom?' and 'appropriate for whom?' That they are controversial, however, in no way invalidates the original questions. Worldwide ecological disruption does exist; social disruption is the norm throughout the world, not the exception. The relative tranquillity of our Canadian society should not blind us to the existence of the social problems which are the lot of the majority of the world's inhabitants.

What is immediately apparent is the profoundly political nature of such questions. Resolving the serious conflicts and basic differences of opinion stimulated by such questions can, in an open society, be addressed only in an open manner through a broad-based political process of public education, public debate, and the reconciliation of conflicting interests. Given the central role of energy in a modern, machine-oriented, industrialized society; the potential for serious social and environmental disruption by man's use of energy; and the seminal role that energy policy has in influencing the pattern and direction of social and economic development,[36] it is clear that the nature of energy policy must be critically assessed in considering such questions.

Today, society stands at a critical time of decisions. The end of the petroleum era forces upon us the requirement to search for new sources of energy. Two distinct energy supply options appear to exist at this time: an option based largely upon nuclear energy and an option based largely upon a variety of renewable energy sources.[37] Both options appear to be feasible, in economic and technical terms (although considerable research and development and institutional organization remain for each before they qualify as 'mature' energy options).[38] However, both options have markedly different implications for society and the environment. It would therefore seem self-evident that the careful intercomparison by energy decision-makers of all the social and environmental costs of these two energy options open to society is essential in choosing an energy strategy which promises to minimize these costs. It would seem essential, in the light of the environmental crisis, that ecological considerations should lie at the heart of all public policy decision-making, including energy policy. To ignore ecological realities in the making of public policy is in the long run tantamount to committing societal suicide.

Unfortunately, the evidence in Ontario is that decisions may already have been taken, which have the effect of committing the province to a major and long-term dependence upon nuclear power.[39] This commitment has apparently been made without a detailed social and environmental cost-benefit assessment of the relative merits of the two differing energy options. Detailed studies comparing nuclear energy with solar energy have not been made; moreover, the commitment was made with no public debate on over-all energy policy, in which the broad, long-term impacts of nuclear energy on society and the environment were identified and discussed.[40]

And yet, the influential role of energy in society and the long-term consequences for the public of energy policy decisions suggest that broad-based public participation in the development of energy policy is essential. In Ontario it is still very much the case that energy policy (at least conceptually) is viewed as a relatively straightforward, technical exercise in balancing a projected demand for energy with energy supply capacity; energy planning decisions generally are taken by a relatively small group of primarily technically oriented individuals. The formulation of energy policy is far from the interdisciplinary process it must be.[41] Recent environmental impact legislation and the action of citizen groups opposed to particular projects have no doubt complicated and slowed the pace at which energy schemes are promulgated. But the attitude on the part of energy planners is very much one of resigned patience: the impact assessment formalities must be gone through and the public given its day in hearings, but then, 'business as usual.' The context in which energy policy must be formulated – a context of powerful vested interests, of great demands upon society's limited resources, and of undeniable costs as well as benefits for society – argues that decision-making with respect to energy policy can no longer be the sole preserve of a small, technically oriented, and largely non-elected (and non-accountable) group of individuals; rather it must be broadened for debate and discussion by much larger numbers of people in a process of a political nature.

A major barrier to be overcome in undertaking such a process is the extent to which the full nature of the environmental crisis and its implications for society remain unrecognized, or, at least, not perceived as imminent, both by the public at large and by its decision-makers. Useful analysis, debate, and action will remain largely impossible until the significance of the problems facing society is widely perceived. This follows from the extent to which the environmental crisis itself originates from the actions of the millions of individuals making up society. Ultimately, changes in attitudes and behaviour will be required at the 'grass roots' level of society.

In Ontario, what may well be required to achieve the necessary large-scale changes in attitude and behaviour is a province-wide, collective education, debate, and decision-making process, on a scale never before attempted by our political institutions, focused on the nature of the problems facing us, the options open in planning future social and economic activity, and the implication of these options. Such a process would need to reach out to every citizen with an urgent request for involvement. It would require a restructuring of the curriculum of the educational system to provide citizens with the information needed to participate in this process. It might well involve the organization of communities and regions of the province into a formal structure for discussion and debate of these issues. Such a structure would, of necessity, interact intensively with our elected parliamentary representatives. Citizen groups, service clubs, trade unions, and industry would need to be involved in a collaborative, interactive, and mutually educating process. This process would have to operate over a long period of time, be flexible, long term in its time-horizon of analysis, and adaptive in order to respond to changing conditions and the evolution of understanding of the issues being considered. It would certainly involve a willingness on the part of our present decision-makers, in both the public and private sectors, to allow a much greater degree of participation by citizens in the provincial decision-making process. It would require substantial amounts of public funds, but probably only a fraction of those required for a single nuclear power plant.

Such a process, although unique in our political history, is not without useful precedents. Indeed, a highly relevant and instructive example exists already in Ontario in the form of the Royal Commission on Electric Power Planning. The chairman of this commission, Dr Arthur Porter, realized from the beginning of his task that the systematic nature of energy in society required far more than a narrow, technical assessment of Ontario Hydro's proposal. Rather, Porter publicly argued that consideration of the merits, or the problems, of the Hydro program could only be assessed through a process of 'inventing the future of Ontario.' That is, only after the province-at-large had answered the question, 'what sort of province do we wish to have in the year 2000?' could the question, 'how much electricity will be required in that future?' be assessed. In attempting to answer these questions, the commission travelled widely, taking its public hearings into many communities throughout the province and establishing an educational and information network in an effort to raise the consciousness of the public. Unfortunately, such a public consciousness-raising, public education, and public involvement process inevitably requires a long period of time, very large amounts of money, skilled manpower, and a major government commitment to provide all these resources. In fact, adequate resources and time were not forthcoming and under pressure from government and Ontario Hydro to complete its task, and faced with budget and manpower constraints, the royal commission quickly retreated into a conventional pattern of public hearings, located in a Toronto office building. The opportunity 'to invent Ontario's future,' and to involve the public in the determination of energy policy, was lost. The 'business as usual' perspective of government and industry carried the day.[42]

Nevertheless a great deal can be learned from the example of this royal commission, and the concept of a province-wide involvement in 'inventing the future' is an exciting one. It is worth noting that there exists in Ontario a very successful example of a broadly based educational and organizational process. Beginning in the 1930s the Farm Forum program in Ontario operated effectively to involve large numbers of farmers in raising markedly the quality of agriculture in the province.[43]

A further suggestive model is provided by the Swedish adult education program, which over a number of years involved many tens of thousands of adults in an examination of the intricacies of the nuclear power issue. This education program is believed to have been a contributing factor to the present re-evaluation by the Swedish government of its nuclear power program. Finally, the Berger inquiry into the proposed Mackenzie River Valley gas pipeline, with its program of taking its inquiry to the people of the region, in a manner and in terms perceived relevant by them, demonstrated how a formal inquiry process can involve effective citizen participation.[44]

The influential role of energy in a modern society, and the great potential for adverse social and environmental impacts emerging from the large-scale production and consumption of energy, suggest that the choice of a new energy strategy to replace petroleum must be a careful one. The growing evidence that a 'business as usual' image of the future may be neither desirable, when its implications are perceived by the majority of our citizens, nor possible, due to a variety of limits and constraints, suggests that it would be an egregious error to allow past energy strategies and institutional commitments, based on an outdated or limited understanding of the role of energy in society and its consequences for the environment, to determine our energy future. Today, in the light of new knowledge, we must reconsider our energy strategy; the message of ecology, the 'critical science,' must pervade our formulation of a new energy strategy; public education and participation – the essence of politics – must be at its heart.

NOTES

1 In this discussion, the term 'environment' is used to encompass the total human environment, made up of both the natural and the social milieux in which man lives.
2 The following provide background reading for the debate on the environmental crisis: P.R. Ehrlich and A.H. Ehrlich, *Population, Resources, Environment* (San Francisco 1972); Ehrlich, Ehrlich, and J.P. Holdren, *World Science* (San Francisco 1976); Report of the Study of Critical Environmental Problems (SCEP), *Man's Impact on the Global Environment* (Cambridge, Mass. 1970); R.L. Heilbroner, *An Inquiry into the Human Prospect* (New York 1974); William Leiss, *The Domination of Nature* (New York 1972); E.P. Eckholm, *Losing Ground: Environmental Stress and World Food Prospects* (New York 1976); T.R. Detwyler, *Man's Impact on Environment* (New York 1971); John A. Livingston, *One Cosmic Instant: A Natural History of Human Arrogance* (Toronto 1973); Lester R. Brown, *The Twenty-Ninth Day: Accommodating*

Human Needs and Numbers to the Earth's Resources (New York 1978); Earl Cook, 'Limits to Exploitation of Nonrenewable Resources,' *Science*, 191 (20 Feb. 1976), 677.

3 D.H. Meadows *et al.*, *The Limits to Growth* (New York 1972)

4 J. Maddox, *The Doomsday Syndrome* (New York 1973); H.S.D. Cole, *et al.*, eds., *Models of Doom: A Critique of 'The Limits to Growth'* (New York 1973)

5 Mihajlo Mesarovic and Eduard Pestel, *Mankind at the Turning Point: The Second Report to the Club of Rome* (New York 1974). The books in note 2 above provide support for the 'Limits to Growth' thesis, as does William Ophuls, *Ecology and the Politics of Scarcity* (San Francisco 1977).

6 Ophuls, *ibid.*, 127

7 In particular, an evolving branch of ecology, which focuses on the totality of man's relationships with his environment, has emerged as an important synthesizing discipline concerned with assembling knowledge from many different disciplines, in both the physical and social sciences, in an attempt to understand the full nature and consequences of this relationship. E.P. Odum, 'The Emergence of Ecology as a New Integrative Discipline,' *Science*, 195 (25 March 1977), 1289

8 See, for example, Ehrlich, Ehrlich, and Holdren, *World Science*

9 Murray Bookchin, *Post-Scarcity Anarchism* (Berkeley 1971), 139

10 M.K. Hubbert, 'The Energy Resources of the Earth,' in his *Energy and Power* (New York 1971)

11 Canada, Energy, Mines and Resources, *An Energy Strategy for Canada: Policies for Self-Reliance* (Ottawa 1976); National Energy Board, *Canadian Oil Supply and Requirements* (Ottawa 1977)

12 Gordon MacNabb, 'Energy: Where Will Canada Be in 1990?' *Geos*, Spring 1977

13 *An Energy Strategy for Canada*, 106

14 D. Pimlott *et al.*, 'Oil under the Ice: Offshore Drilling in the Canadian Arctic,' Canadian Arctic Resources Committee (Ottawa 1976). The newsletter of CARC provides further information on the existing and potential ecological impacts of northern Arctic oil and gas development.

15 W.J. Campbell and S. Martin, 'Oil and Ice in the Arctic Ocean: Possible Large-Scale Interactions,' *Science*, 181 (6 July 1973), 56

16 J. Macpherson and Greg Thompson, 'Polar Gas: A Premature Pipeline?' *Alternatives*, 7, no 4 (Autumn 1978), 34

17 Thomas R. Berger, *Northern Frontier, Northern Homeland: The Report of the Mackenzie Valley Pipeline Inquiry* (Ottawa 1977), I

18 B. Christiansen and T.H. Clack, 'A Western Perspective on Energy: A Plea for Rational Energy Planning,' *Science*, 194 (5 Nov. 1976), 578; Boyce Richardson, *James Bay* (Toronto 1972)

19 On this problem, see Report of the Study of Man's Impact on Climate (SMIC), *Inadvertent Climate Modification* (Cambridge, Mass. 1971); S.H. Schneider, *The Genesis Strategy: Climate and Global Survival* (New York 1976); Alvin M. Weinberg, 'Global Effects of Man's Production of Energy,' *Science*, 16 (1974).

20 Edward Goldsmith, 'The Future of an Affluent Society – The Case of Canada,' *Ecologist*, 7, no 5 (June 1977), 160

21 The intercepting of a relatively small portion of the incident solar radiation, its utilization for man's purposes and subsequent release to the environment as heat, produces neither air pollution nor a significant net increase in the heat content of the atmosphere. This may be compared with the much more serious environmental impacts of

carbon-based energy sources. A recent review by Stan Terra, 'CO_2 and Spaceship Earth,' in the *Electric Power Research Institute Journal*, 3, no 6 (July–Aug. 1978), 22, quoted a senior research climatologist with the National Oceanic and Atmospheric Administration as saying: 'ours is the generation that must come to grips with the carbon dioxide problem ... There is almost no aspect of national and international policy that can remain unaffected by the prospect of global climate change. Carbon dioxide, until now an apparently innocuous trace gas in the atmosphere, may be moving rapidly toward a central role as a major threat to the present world order.' The article calls for the development of non-carbon based energy sources for worldwide use in both industrialized and underdeveloped nations.

22 *An Energy Strategy for Canada*, 108, 68
23 Dillon, 'Ontario – Towards a Nuclear Electric Society?' paper presented to the Canadian National Energy Forum, 16 Oct. 1975
24 Ontario Hydro, *Long-Range Planning of the Electric Power System*, Report no 556SP (1 Feb. 1974); this original report was updated by Hydro report LRF48A in Feb. 1977.
25 The royal commission, under the chairmanship of Dr Arthur Porter, was instructed to examine the long-range electric power planning concepts of Ontario Hydro for the period 1983–93 and beyond, to relate them to provincial planning, to the utilization of electrical energy, and to environmental, energy, and socio-economic factors. A very large, often contradictory, and sometimes contentious literature exists as to the costs and benefits of nuclear power programs. Representative of the supporters of nuclear power is the article by H.A. Bethe, 'The Necessity of Fission Power,' *Scientific American*, 234 (Jan. 1976), 21. A very detailed critique of nuclear power is that by Amory Lovins, *Non-Nuclear Futures: The Case for an Ethical Energy Strategy* (Cambridge, Mass. 1975). A description of Canadian nuclear technology is contained in an article by J.A.L. Robertson, 'The CANDU Reactor System: An Appropriate Technology,' *Science*, 199 (10 Feb. 1978), 657. A Canadian critique of nuclear power, 'Half Life: Nuclear Power and Future Society,' has been prepared by the Ontario Coalition for Nuclear Responsibility.
26 Ontario, Ministry of the Environment, Submission to the Royal Commission on Electric Power Planning, *Final Hearings*, V: *Nuclear Power*; Lovins, *ibid.*; J.W. Gofman and Arthur R. Tamplin, 'Low Dose Radiation and Cancer,' *IEEE Transactions on Nuclear Science*, NS-17 no 1 (Feb. 1970), 1–9; Mason Willrich and T.B. Taylor, *Nuclear Theft: Risks and Safeguards* (Cambridge, Mass. 1974). The government has been sufficiently concerned with the economic and political implications of Ontario Hydro's expansion plans to have two select committees scrutinize these plans, simultaneously with the royal commission hearings. See, for example, 'A New Public Policy Direction for Ontario Hydro: The Final Report of the Select Committee of the Legislature Investigating Ontario Hydro,' June 1976.
27 Ministry of the Environment, *ibid*.
28 One of the most detailed comparisons of the nuclear power option with the solar energy option is that by Amory Lovins, *Soft Energy Paths: Toward a Durable Peace* (London 1977); see also Middleton Associates, 'Canada's Renewable Energy Resources: An Assessment of Potential,' study prepared for the Office of Energy Research and Development, Department of Energy, Mines and Resources, Canada, 1976.
29 Wilson Clark, *Energy for Survival: The Alternative to Extinction* (Garden City, NY 1975); Ministry of Energy for Ontario, 'Turn on the sun,' 1977
30 D. MacKay and R. Sutherland, 'Methanol in Ontario: A Preliminary Report,' Ministry

of Energy for Ontario, 1976. See also: Peter Love and R. Overend, 'Tree Power: An Assessment of the Energy Potential of Forest Biomass in Canada,' Report ER-78-1, Canada, Energy, Mines and Resources, 1978.

31 For a discussion of federal energy minister Alastair Gillespie's five-year $380 million renewable energy program, see *Canadian Renewable Energy News*, 1, no 8 (Aug. 1978).

32 This approximate energy delivery equality excludes consideration of electricity produced from renewable hydroelectric sources and is based on federal government estimates (*An Energy Strategy For Canada*, 104), and a conversion to heat of nuclear generated electricity of 3412 BTU s per kwh

33 A. Lovins, 'Exploring Energy Efficient Futures for Canada,' Science Council of Canada, *Conserver Society Notes*, 1, no 4 (May–June 1976); David B. Brooks, 'A Real Option: Conservation to 1990 and Beyond,' *Alternatives*, 7, no 1 (Fall 1977), 48; P.A. Victor, 'Alternatives to Ontario Hydro's Generation Program,' report prepared for the RCEPP by Middleton Associates, Nov. 1977

34 'Solar Thermal Test Facility,' *Sky and Telescope*, 55, no 4 (April 1978), 286

35 A growing body of literature examines these questions: Edward Goldsmith *et al.*, 'A Blueprint for Survival,' *Ecologist*, 2, no 1 (1972), 1; for a Canadian 'Blueprint,' see Canadian Survival Institute of Canada *Draft Canadian Plan for Survival* (Toronto 1974); Denis Hayes, *Rays of Hope: The Transition to a Post-Petroleum World* (New York 1977); E.F. Schumacher, *Small Is Beautiful: Economics As If People Mattered* (New York 1973); Garrett Hardin, *Exploring New Ethics for Survival: The Voyage of the Spaceship Beagle* (New York 1972); J.P. Holdren and P.R. Ehrlich, eds., *Global Ecology: Readings Toward a Rational Strategy for Man* (New York 1971); Robert Disch, ed., *The Ecological Conscience: Values for Survival* (Englewood Cliffs, NJ 1970); 'Canada as a Conserver Society: Resource Uncertainties and the Need for New Technologies,' Science Council of Canada report no 27 (Sept. 1977); P. Shepard and Daniel McKinley, eds., *The Subversive Science: Essays Toward an Ecology of Man* (Boston 1969).

36 For example, a decision to continue to meet a growth rate in demand for electricity of 5 per cent per year will inevitably require the construction of large numbers of nuclear power plants, is likely to reinforce traditional patterns of economic growth, and will have a particular set of implications for society and the environment. On the other hand, a decision to emphasize energy conservation and the substitution of renewable energy technologies for electrical generation facilities would tend to encourage a different pattern of economic activity with, arguably, a substantially different set of implications for society and the environment.

37 Clearly, any real world energy strategy involves a mix of energy forms. For example, the pursuit of a renewable energy policy in Ontario would not necessarily involve the immediate shutting down of all nuclear power plants. Even in a solar energy future, there may well be a carefully designed role for nuclear energy. In speaking of nuclear or solar options, we are talking about the emphasis given to these energy sources in energy policy.

38 In the case of nuclear power, this includes the development of waste disposal methods, fuel recycling methods, a new thorium-based fuel cycle (see Robertson, 'The CANDU Reactor System'), methods for dealing with growing public opposition, and, a recent development, measures for dealing with the stretching of pressure tubes in reactors under the intense bombardment of neutrons. *Globe and Mail*, Toronto, 'Problems in

Reactors Likely to Cost $500 million,' 15 Aug. 1978. In solar space heating applications, for example, solar collectors, which function efficiently in Canada's cold winters, are long-lived, and are cost effective, need to be made commercially available; large-scale manufacturing, marketing, installing, and servicing institutions need to be developed; 'rights to light' legislation may be important.

39 Jonathan Thorpe, 'Ontario Opts for Nuclear Power,' *Toronto Star*, 30 May 1977. The closest that the Ministry of Energy has come to issuing a long-range energy policy document is 'Ontario's Energy Future,' 1977, which clearly favours the nuclear electric option while stating that renewables will be unable to contribute significantly to provincial energy requirements until well into the next century.

40 The RCEPP is nominally undertaking just such an assessment; however, it is very much facing a *de facto* situation. Already a large commitment – in the form of institutions, large numbers of people involved in the nuclear industry, prestige, resources, and time – has been made to the nuclear future.

41 'Canadians Denied Voice on Energy Policy, OECD Says,' *Globe and Mail*, 29 June 1978. This report, which describes a 200-page draft OECD report, quotes the author as saying that the Atomic Energy Control Board (the supervisory agency for nuclear matters in Canada) has operated for most of its thirty years 'outside the forum of major public debates and within the narrow world of the nuclear industry and electric utilities.'

42 Naturally, many of the most fervent supporters of Ontario Hydro's expansion plans are industries dependent on large amounts of relatively inexpensive, dependable electricity. Nuclear power with its ability to provide large blocks of power allows traditional patterns of economic (and energy) growth and development, leading to a vision of tomorrow very much the same as today, only 'bigger.' On the other hand, an energy policy which emphasizes energy conservation and the harnessing of our renewable energy resources, at the very least, will require some modest innovation in these traditional patterns of growth; at worst, such an unknown energy strategy might be perceived as downright threatening to existing institutions and industries.

43 Rodger Schwass, 'Canada's Farm Radio Forum,' unpublished PHD thesis, University of Toronto, 1971

44 A good description of Swedish energy policy is contained in the proceedings of a Swedish Energy Policy Seminar, arranged by the Swedish Consulate, Toronto, 1977; Berger, *Northern Frontier, Northern Homeland*

10

Southern Manitoba: an experiment in regional action

J.E. Page and M.E. Carvalho

Many times has the story been told of the characteristic settlement pattern for the Canadian prairies: the demands of steam engine technology combining with the daytime hauling distance of a horse-drawn wagon to determine the location of thousands of settlements across western Canada as hundreds of thousands of people trekked out onto the prairies in the late nineteenth and early twentieth centuries. What is generally overlooked in these accounts is the 'original report' on settlement of the Canadian prairies which was submitted by Captain John Palliser to the British government in 1862.[1] The questions the British government was investigating then were not unlike the ones we are concerned with today – is this prairie region of Canada habitable, and, if it is, to what extent, and under what assumptions about environmental conditions?

The span of a century has added little to the insights of decision-makers still faced with this question. Palliser noted forcefully that there was great hazard in settling the prairie region of Canada; only on the narrow strip of black-soiled territory at the edge of the boreal forest, where precipitation was fairly reliable, was it safe to consider permanent settlement. However, misguided experiences with cropping the prairies just after the transcontinental railway went through two decades later created the impression that the whole prairie region was a vast fertile territory awaiting only the invention of the deep plough to produce boundless amounts of grain for a planet with an exploding human population.

The setting for this account of an experiment in 'radical regional action' encompasses about 75,000 square miles (including about 12,000 square miles of water surface) of southern Manitoba. The area is bounded on the north by the 53rd parallel, on the south by the 49th parallel (the United States border), on the east and west by the Ontario and Saskatchewan borders respectively. The study area excluded directly the provincial capital: Winnipeg the city had for a long time been the focus of attention by all three levels of government (municipal, metropolitan, provincial) at the cost of adequate attention to the needs of rural Manitoba.[2] Although it was realized at the very outset that an action study program which was not focusing on Winnipeg would be wanting in some respects, it was felt that the exclusion of the large urban area would permit a more concentrated identification and assessment of conditions characteristic of rural southern Manitoba, in a way that would lead to widespread participation by the residents of rural Manitoba.

The study area of the Regional Analysis Program[3] (RAP) offers a striking variety of environmental, economic, and social features. A large percentage of the land is fertile, but there are areas of uneven fertility and unpredictable precipitation; there is a combination of prairie grass-land, boreal forest, and Canadian Shield ecosystems; there are numerous shallow bodies of water and countless drainage problems; finally there are climatic uncertainties and extremes that have challenged the process of settlement and development for more than a century and a half.

The pattern of settlement in southern Manitoba was largely determined by the early subdivision of townships, sections, and quarter-sections,[4] and road allowances overlaid by an elaborate network of railway lines reaching out to gather in the grain harvests and to distribute manufactures. This pattern has today resulted in more than 300 settlements, ranging in size from 50 to 31,150 people, with well over 80 per cent of them having less than a thousand inhabitants. Over-all, the settlements, plus farms and Indian reserves, provided a population for the study area of about 400,000 people in 1971.

The urban hierarchy in southern Manitoba has been termed extraordinarily 'imbalanced': its metropolitan centre, Winnipeg, has a population of about 540,260 (1971); Brandon, 140 miles directly west of Winnipeg, is the next largest city with a modest population of only 31,150; and there are four other centres with populations of 5000 to 13,000. This 'imbalance' is testimony to one of the most important problems with which the Manitoba government (and hence RAP) has been concerned: that of rural out-migration as part of the process of urbanization in the prairies. This process, in essence, has resulted in rural depopulation along with an ever-increasing dominance of Winnipeg in the settlement system. For example, between the years 1951 and 1971 the rural population of the region declined by over 13 per cent while the 'urban population'[5] increased by over 41 per cent. Likewise the number of farms decreased during the same period by about 32 per cent, while average farm size increased by 58 per cent.[6] With the decline in the farm economy (it seems that no notable increase of income results from the larger farming operations), there has been a fall-off in rural services and an increase of political apathy; the future of rural Manitoba did indeed appear bleak in 1971. Action was needed – but where was it to start and how was it to proceed?

RAP AS A MEANS FOR RADICAL SOCIAL CHANGE

It was apparent from the outset, taking into account both ecological and settlement conditions, that nothing short of 'radical social change' was needed in southern Manitoba. This meant that both interests and grievances had to be identified and addressed, that a willingness on the part of the people to enter into this change process had to be cultivated, that the people had to be convinced that they could be instrumental in altering the direction of change on their own behalf. The direction of environmental rundown over the century of settlement since Palliser's report was not the result of fate; it was the outcome of decisions made by people somewhere. A century of living in this section of the Canadian prairies clearly indicated that the process going on was neither benefiting the natural ecosystem nor the people settled on the land and in small settlements.

When it is said that RAP sought to bring about 'social change' it is meant that the consciousness of the people of the region had to be altered, a new awareness of their own lives and of the environmental setting for these lives had to be cultivated; a new self-perception had to develop. The evolution of this self-perception inevitably involved a high degree of community learning. Planning was seen to involve a process of learning for regional self-determination. RAP as a staff research planning group was seen not as a new government agency but simply as a catalytic redirecting influence in the process of on-going change. Information development and flow was seen as being fundamental in grappling with this modification. It was assumed that information would allow decisions at all levels to be made with both a wide and a focused interest, so that response would take on a positive note. Planning would be used *in* the change process rather than *on* the change process in a cyclical manner: 'change – challenge – research – decision – action response – change' was the sequence that needed to be deliberately cultivated in a conscious manner. Change itself was seen to be inevitable, but not the quality of the change: that was 'up for grabs'! And it was clear both ecologically and socially that the quality of modification had to be altered in a manner better suited to both the over-all life-support system and the social and economic conditions for human living.

Previous efforts to improve living in southern Manitoba had emphasized economic change, research for new industries, development of agribusiness, enlargement of the service sector, etc. RAP was directly concerned with the quality of life in the small urban centres, and it sought a method for relating one centre to another so that together they would be able to provide a higher level of service. There was even the prospect of ultimately redesigning the municipal and provincial governmental structures, some shifting of the centre of dominance away from Winnipeg, so that genuine self-support structures and processes could develop throughout rural Manitoba. This was not to be done by way of conventional planning *for* the people of the area. Rather it had to evolve from the interests and convictions of the people; their ingenuity, insight, effort, and experiences had to be a starting point. Quite obviously this kind of change could not be forced or hurried; it had to be evolutionary.

One of the main assignments given RAP was to identify community interaction patterns; it was seen to be basic for government policy to understand just how one settlement related to another, especially when trends in other prairie areas suggested that the time had come to consider letting some settlements die while others would be encouraged as growth centres. An earlier experience with this kind of concern clearly taught the RAP advisers that the determination of the people to survive and prosper can be the critical factor as to whether settlements die or prosper. On the other hand, it was clear from the history of settlements in the prairies that little concern had been given to ecosystems in establishing settlement location and activity networks; all was conceived as a function of steam locomotive technology. And yet, no thought-out readjustments had been made since the days of early settlement. Population decrease, depletion of the soil fertility, drought, isolation, and marginal social life were really symptomatic of the need for radical change. And yet radical change too quickly introduced could cause needless resistance to change even though it was clearly appropriate for the problems at hand.

When radical social change is considered in RAP it cannot be thought of separately from change in the natural ecosystem. For a century, the Manitoba prairies, like the rest of the western prairies, were really envisaged in factory terms; farming, or better, 'grain mining,' was really a factory operation. The living conditions and the work conditions of the workers were of little concern to the grain market system; what mattered was the production of grain for export. The scattering of the people over the prairies by the quarter-section system was the clearest evidence of the factory model at work. Except in those few locations where ethnic conditions dominated there was no concern for the cultural life of the people. It was a matter of grim, harsh survival accompanied by uncertainty of outcome for all the effort expended. Ironically since the end of the war in 1945 there has been a notable increase in farm size. It was assumed that larger crops could be taken off and a higher level of farmer prosperity attained. Again the anology of the factory or of the mine is appropriate. The information turned up by RAP indicated that farmer prosperity had not been increased by larger farm size nor has farm fertility. Improved machinery allows for more 'efficient' cropping. Chemical fertilizers sustain the yield. But input costs have been increasing, basically, a fact of life brought out clearly only with the energy crisis. So long as energy costs were low, the actual cost of input through increase of technology and chemical fertilizer was not heeded. The evidence for all this shows up in the statistical data from RAP: despite the 32 per cent decrease in the number of farms and the 58 per cent increase in the size of the farms, the farm income per unit showed no substantial change in real income, if one allows for cost of living increases.

To give some idea of how 'radical' the approach of RAP has been to the problems of southern Manitoba one can look at the situation of the native poeples. Consultations were held with the leaders of the native and Métis peoples at the very outset of RAP in order to assure their co-operation and to obtain their approval of the information that was to be made public about native peoples. For the first time both the statistical and map plate presentations of data showed clearly just how the native peoples were serviced with respect to other sectors of the population. As well, the demographic information so long buried in over-all aggregations of population counts was taken apart and reassembled so that the real make-up of the native peoples' settlements would be known.

Besides the new detailed information a special set of programs for native peoples' assistance was worked up and supplied to the Rural Region Working Group (which directed RAP) through a special report. To illustrate the approach, consider the question of housing for native people. The provision of housing must not start from any of the conventional housing models but with the culture of the bands in any particular area. The problem is not viewed as one of supplying houses but rather of 'housing' native people, that is, of engaging the people in the design of the housing, siting of the housing, technical details of construction, etc. Integral to such an approach is an in-service training program that allows the people both to acquire building trade skills and to see how their homes are put together so that as time goes on and maintenance has to be done they will have the requisite skills to carry it out effectively.

Not only construction but financial aspects of the housing activity have to be worked out in terms of the income and savings plans that various bands may have attained. The food-gathering habits and the traditions of cooking and food preparation need all to be taken into account so that the house is fitted to their practices. In this way the option is kept open that favours the preservation of aspects of native cultures. The intricate process of dealing with federal and provincial housing officials opens up new opportunities for job training and continuing work.

The whole housing activity program is cast within the context of a 'clustering principle': while one or even two native peoples' settlements may not provide enough long-term demand for a small housing industry, by their combining of construction and maintenance work in three, four, or five settlements a new industry could be developed which would not only supply services to the people but would call upon them to view their natural environment in a new light, because from it they would be drawing most of the renewable resources for building and maintenance material.

RAP AS A RESPONSE TO ECOLOGICAL CONDITIONS

The RAP model is grounded in a modification of a scheme developed by Sir Patrick Geddes at the turn of the century in Britain. The modification incorporates systems terminology in order to highlight the dynamism of the 'place/folk/work'[7] basis of Geddes' synoptic approach to regional planning. As the diagram of the model below indicates, the simple trilogy of Geddes has become: environmental sub-system; socio-demographic sub-system; activity sub-system. Each sub-system was defined as being composed of variates capable of description and analysis (for example, community services, utilities, soil characteristics, housing, transportation)

This kind of model served well to provide both a starting point for information gathering and a means of controlling the vast amount of information once the program was underway. Just as the map displays were later intended to do, the modelling in this fashion facilitated the development of an increasingly enlightened synoptic view of southern Manitoba. It was intended to forward both synthesis and analysis.

At the outset it was mentioned that from the time of first white settlement on the prairies there has been degeneration of environmental conditions and a failure to build settlements that serve to enrich human culture. The setting down of three sub-systems in the model on an equal basis made explicit the intent of the RAP effort to soften the dominance of the economic considerations that marked previous efforts in the province to improve living conditions – Agricultural Rural Development Act (ARDA), Committee on Manitoba's Economic Future (COMEF), and Targets for Economic Development (TED) – and to update the social and ecological aspects of settlement. The latter are seen as 'life needs' and are intertwined; if the natural environment is in a state of decline, then the human conditions for living are in decline. Social and ecological aspects are linked through the 'activities sub-system' in the model to emphasize where both cause and remedy to both human and ecological sub-systems' decline or uplift are located.

How could RAP respond to the demands of ecological conditions in southern Manitoba? It was clearly impossible to mount a separate study of ecological conditions as it was possible to do on a sample basis for social concerns. Fortunately it was possible to generate a high level of co-operation between the local Canada Land Inventory (CLI) staff and RAP staff. Nearly all the inventory of the 70,000 square miles of RAP concern was completed by the Canada Land Inventory. Nevertheless, uncontrollable circumstances resulted in a two-year delay between the publication of the first RAP map plates and the final RAP land capability maps. This was a critical delay in the light of the intent of the RAP model to present with a measure of equal concern the demographic and the economic and environmental information.

This overview of the environment for the RAP region was technically difficult to deal with and communicate about. In fact, the working paper on the environment is still in production. Part of this difficulty comes from the scale at which the information is presented and part from the classification principles of the CLI. Much has been learned since CLI began and new classifications for many uses of land are needed. The very great importance of marshes and deltas, the sensitivity of sandy shorelines of lakes, the dominating importance of water quality and water management where uncertainty of climate and precipitation prevail, all conspire to demand of the CLI a level of detail not really possible without great costs. And yet the highly complex system of overlays produced by RAP do bring out an immensely important fact about the southern Manitoba area: there is great variety in the natural system, and much local knowledge is required to deal effectively with the ecosystem in a way that will increase the value of the natural inheritance over time instead of continuously eroding its capacity for human and other life support.

The RAP was able to develop a new method of handling the CLI materials called 'Composite Natural Environment Capability Maps.' Operating still at the scale of the other map plates (1:800,000), RAP is able to offer the kind of information needed by rural residents and users of the natural environment to discover some of the impact of policies and programs that optimize man's use of the environment while at the same time minimizing any adverse effects human activities might have. This method of presentation highlights the variety of uses any parcel of land may be put to. At the same time the range of choices points up the complexity of the southern Manitoba environment and the need to examine carefully any activities proposed or already being carried out. That is, the RAP composite maps allow a qualitative assessment of compatibility to be made between 'use' capability and existing land uses, an all-important first step in rethinking the whole settlement and land-use inheritance that characterizes southern Manitoba.

THE REGIONAL ANALYSIS PROGRAM MODEL

The accompanying diagram provides a schematic outline of the model developed by the authors in conjunction with the Department of Industry and Commerce. It evolved out of a pilot model of 1970 applied to a 20,000 square mile section of the eastern part of southern Manitoba. Practically all the essential elements of the model had been tested through this earlier smaller-scale effort, so that it was brought forward with a measure of confidence in 1971 for larger-scale testing in a territory over three times the area of the 1970 application and with a population six times larger. As the diagram shows it is the parallelism of the model that is noteworthy. On either side, *equal* importance is given in both structure and process to the evaluation of data by local communities and by government. It was this emphasis which found expression through the process of Community Committee Reporting in 1972. The model

A regional analysis & planning process model

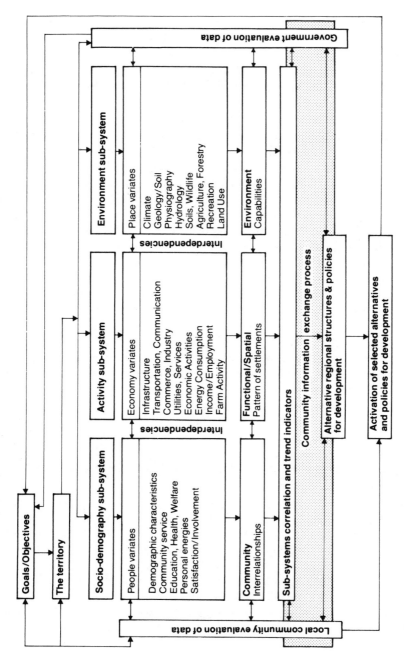

Goals/Objectives

The territory

Socio-demography sub-system

People variates

Demographic characteristics
Community service
Education, Health, Welfare
Personal energies
Satisfaction/Involvement

Community
Interrelationships

Sub-systems correlation and trend indicators

Interdependencies

Activity sub-system

Economy variates

Infrastructure
Transportation, Communication
Commerce, Industry
Utilities, Services
Economic Activities
Energy Consumption
Income/Employment
Farm Activity

Functional/Spatial
Pattern of settlements

Interdependencies

Environment sub-system

Place variates

Climate
Geology/Soil
Physiography
Hydrology
Soils, Wildlife
Agriculture, Forestry
Recreation
Land Use

Environment
Capabilities

Government evaluation of data

Local community evaluation of data

Community information exchange process

Alternative regional structures & policies for development

Activation of selected alternatives and policies for development

Regional Analysis Program
Carvalho/Page Group   1972

also indicates that the inevitable outcome of this dual evaluation process would be the encouragement of dialogue between local residents and the government. The horizontal arrows between the sub-systems indicate the *interactive* nature of both information and decisions, decisions and actions. Action engendered in one sub-system is seen to call up action and reaction in another.

Reading from top to bottom provides a key to the production of the map overlays[8] with an ever-increasing amount of related information being brought together on individual sheets. Each topic listed had at least one map display showing basic relationships. This was intended to move the awareness of people to the radical complexity of the issues and of the need for co-operative responses to the expression of needs from the people once they had the advantage of new information about their own area and of the region as a whole. The return direction of the arrows in the system line-flows indicates that action does not wait for the unreal expectation of complete information; rather it is based on sufficiently worked-out information which will itself generate not only more needed information but also a deeper grasp of related information already gathered for the sake of both understanding and action.

The diagram of the model clearly highlights the kind of 'goal' aspect used to confront the change process in southern Manitoba. While 'goals' are indicated at the top of the diagram they are not conspicuous in actual operation. An overriding strategy determined to keep the goals implicit; only at first was it mentioned that fundamental social change was the goal of the whole process. Given the nature of the model, goals could not really be handled in any other fashion; they required a level of knowledge and consciousness about the region and its people and activities that was simply impossible to possess at the beginning of the RAP process. Goals for each sub-system had to be sought independently and within a synoptic view of the region.

There were two distinctions made at the outset: goals of the RAP process and goals for the southern Manitoba region. The former were developed by the designers of the RAP process; the latter had to be discovered by the people of southern Manitoba individually and collectively, in their public and private institutional groupings. It was the goal of the RAP process that the goals for social change in southern Manitoba be discovered as the RAP process unfolded. The vast informational base which was to be built up and kept updated[9] was to be a support element in goal discovery for the evolving reality of southern Manitoba.

PRINCIPLES OF THE REGIONAL ANALYSIS PROGRAM

The approach of RAP to social change and ecological conditions was grounded in the following principles:

1 / If participation by citizens is to be effective it must be grounded in a sound information base; consultation without appropriate information is unsound.

2 / Regional development is not a program *for* people; it is a program involving the people of an area in deciding upon and working to improve their region (and inevitably altering themselves in the process).

3 / Unless environmental realities are dealt with at the outset, decisions will be made for economic gain which ultimately destroy natural resources to the detriment of long-term economic and social benefit.

4 / Sound regional planning/development must be grounded in a people who plan to stay on the land and are concerned about long-term settlement and the improvement of the quality of living in settlements over time.

5 / Leadership and entrepreneurial activities are as important to encourage as are government initiatives in any regional scheme for over-all improvement.

These principles evolved through the experience of the East-Man study of 1970. They came to coincide closely with the 'Four Principles' of *Guidelines for the Seventies*, an official government statement that 'forms the basis for a dynamic flexible plan designed to be responsive to the requirements of the people of Manitoba – a plan whose program components can be evaluated in terms of basic development policy principles and can be adjusted according to the wishes and changing needs of our society.'

The basic principles as set out in the *Guidelines* are:

'1 / maximization of the general well being of all Manitobans;

2 / greater equality of the human condition for all Manitobans through a more equitable distribution of the benefits of development;

3 / implementation of an effective stay option through policies and programs which will prevent Manitobans from being coerced by economic forces to leave their province or to leave the region within the province in which they prefer to live; and

4 / the promotion of public participation in the process of government, and more particularly, in the development decisions which will affect all Manitobans in the years ahead.'[10]

As can be seen, the *Guidelines'* principles reflect a different attitude towards the natural environment from that of the RAP principles. Clearly a consensus on environmental matters was not possible at that time in the government.

THE RAP PROCESS

The Regional Analysis Program did not begin in a vacuum. It built upon earlier recommendations for action to improve living conditions in Manitoba based on commissioned studies undertaken by the Conservative government during the sixties before the New Democratic party took power in 1969. The first of these, the Committee on Manitoba's Economic Future (COMEF), and the second, Targets for Economic Development (TED), strongly emphasized the economic side of the efforts to deal with problems of survival and development in southern Manitoba. However, they both agreed on the importance of establishing and maintaining a set of institutions known as 'Regional Development Corporations.' They were intended to forward planning and development interests in each of their assigned territories: eastern Manitoba, Interlake, Pembina Valley, Central Plains, western Manitoba, and Parklands in the northwest. As voluntary but official organizations jointly supported by local government and business and the provincial government, they were essentially intended to be a new linkage between the local areas and the provincial government. The appropriateness of their functions in the light of the RAP exercise led to RAP involving the corporations as much as possible even though there was uncertainty about their continuance at the time by the new government. In fact they played an effective role in the RAP efforts.

RAP went forward in a series of 'phases' not necessarily in a linear sequence but in an interacting fashion. Thus:

phase I questionnaire survey, information gathering, organization and publication of information giving a basic overview of southern Manitoba as a region. Standards of data presentation and classification were set so that as the program proceeded greater clarification could be gained in the process of studying community interaction with comparative data.

phase II an analysis/synthesis stage during which a non-interpretive series of reports and map presentations of information was prepared and published in order to build up an increasingly integrated view of the people, their activities, and their environmental setting in southern Manitoba.

phase III phases I and II are really for the sake of phase III, the information exchange phase, during which the information was extensively shared with the people of southern Manitoba. During this phase they were to study the information and report back to the government on how they understood their own communities and their needs.

phase IV involved preparation of innovative responses to the demands generated in phase III in order to have practical and effective testing of the capacity of government and people to work together in the common desire to improve the quality of living in southern Manitoba.

phase V this phase was seen to go forward on the basis of successes and failures in phase IV. The hope was that there would be integrated responses, co-ordinated action on the part of line departments working together in response to demands from citizen expressions of need. As well it was expected that individual initiatives and local leadership at both community and Regional Development Corporation levels would begin to operate with growing experience and confidence in a new style of planning/development for the parts of the region as well as the whole region.

Phases I and II: information gathering and organizing
At the outset of RAP it was agreed that 'community interaction patterns' were one of the most important gains to be obtained from the first phases of the study. To accomplish this a detailed questionnaire was designed and administered to a 5 per cent sample of the 100,000 families of southern Manitoba. This provided up-to-date information to be combined with older data already gathered and organized. Two classes of outputs were provided: graphical and statistical. The first employed a map overlay method in order to assist anyone using it to increase his or her ability to grasp relationships between the information being presented on any individual sheet and another sheet, or to relate information in any series of sheets. For both methods of presentation, loose-leaf and folio formats were used to assure ease of keeping the material updated.

The information gathering as such, both formally and informally, from government departments and private and public agencies, was intended to be a first step in building up the information exchange process immediately, and later in carrying out co-operative activity. In fact it was discovered that (*a*) there was an immense amount of relevant information available and, as it were, waiting to be used, and (*b*) that much good will could be engendered for the program by this demonstrated interest in work done by others over many years and who up until this time saw their efforts largely wasted. At the same time it was mandatory that expectations would not be roused for any specific kind of government responses to needs except those that would be generated by the combined efforts of all parties as the program moved forward.

Basic field work began on 1 May 1971 and extended to the end of August. This timing was influenced by the availability of students and the critical need in 1971 to generate jobs for university students. At the peak of activity as many as forty students were involved in all phases of the work. While the central office was located in Winnipeg, the experience of the previous year with the East-Man study demonstrated how important it was that the offices of the six Regional Development Corporations (RDC) be used as much as possible and that the people and the corporations come to realize that there was to be a real but gradual decentralization of responsibility for planning change and development as the program evolved.

The first round of information gathering and survey work was substantially completed by early fall of 1971; by December final drafts of much of the graphical and statistical information were ready for printing.

Phase III: information exchange
During the fall of 1971 it was decided that two 'caravans' would travel through rural Manitoba in order to bring the information outputs directly to as many people as possible. Two line departments of the provincial government co-operated in the budgetting and organization of this phase. Approximately seventy-five committees (one for each settlement with 500 or more population) were established under the aegis of the Regional Development Corporations and with the assistance of staff from both the Department of Agriculture and the Department of Industry and Commerce (the latter initiated and administered RAP).

A critical step in this organizing process was a three-day training session for representatives of each community committee. Emphasis was placed on having delegates under thirty years of age attend this seminar.[11] Its purpose was to explain the information which had been organized to date and what was being attempted throughout southern Manitoba. It was also to be the beginning of training sessions for those persons who would accompany the caravans which were to travel from the third week of January and to the end of March 1972. Following a caravan visit, each community committee was assumed to be able to make effective use of the array of published information from the program.[12]

During the summer of 1972 the principal efforts of the RAP staff centred on analysing the data gathered to date and reducing it to report form for those working on the development of community statements of need for improvement in the region. The response of the community committees was greater than expected and produced four two-inch thick volumes of statements from some eighty-five working committees. All this material was then gone through first at the RDC level and then at the central RAP office to reduce it to workable size for transmittal to and consideration by a cabinet committee.

By the end of the summer 1972, summary statements embodying the results of their respective community reports were produced by the six Regional Development Corporations. While weaknesses were perceived both in the procedures used by some committees and in the reports which they produced, the over-all results indicated a remarkably sound and widespread support for the efforts being made through RAP. During the fall and the early winter of 1972, statements of need in the reports were organized and by the end of January and early February 1973 a series of meetings between selected cabinet ministers and the community representatives in each RDC area was held to discuss ways and means of responding to these expressed needs.[13]

Phase IV: innovative responses
Concurrent with the efforts to integrate the results of the community committees and RDC reports and a first round of discussion with the ministers, there was a reorganization of the steering committee of cabinet ministers which had been helping in the direction of RAP. Experience gained from steering committee meetings and special study days during the first half of 1972 indicated that structural changes ought to be made in the way of directing and managing RAP. The steering committee was reconstituted by the Planning and Priorities Committee of cabinet as the 'Rural Region Working Group.' Its principal task was to carry forward the preparation of 'clusters of solutions,' establish 'lead agencies' to carry forward these solutions, and generally attempt to effect practical responses to the expressed needs of the people of southern Manitoba.[14]

The notion of responding in 'clusters' suggested that in fact there are not individual settlements getting by on their own in southern Manitoba. Rather, there was discovered to be groupings of settlements which function and interact together in mutual interdependence. By acknowledging and taking advantage of this interdependence it was assumed that a multiplier effect could be achieved with individual projects that would be undertaken at the local or regional level. That is, greater benefits with less cost were anticipated.

Phase V
As mentioned earlier, the 'phases' were not intended to be linear in sequence. In fact, phase III was well advanced before phase II was far underway; Phase IV was underway before phases I and II were completed because of the need to keep the information up to date and because of new information needs being developed as the program moved ahead generally. In other words, it was not assumed that the data gathering and organizing and presentation would ever stop once the planning/development got under way.

In actual practice it may be incorrect to identify phase V as part of the program because it was expected that the 'five phases routine' would soon integrate into the regular functioning of government. That is, it was the intention of the program to try to improve the communication process for sake of response to change within the operating system and not to produce another separate agency within the government service. The additional operating units would be the 'community committees' which on a volunteer basis could provide the basis for the people of the communities to have access to the information needed (as the operating of RAP indicated) and would be a liaison group between the Regional Development Corporations and the communities (and, if necessary, a liaison between the communities and the provincial government departments).

Examination of the content of many expressed needs set down by the community committees in phase III showed that many of these needs could best be attended to by the people themselves, with or without some minor assistance from the government. Other needs could be attended to largely by the line departments with very little modification of present practice and with present resources. The most important alteration might be the integration of the responses in closer co-operation with the efforts of other departments to make the total response far more beneficial for both communities and the larger region. A third class of responses clearly called for 'innovative solutions' to the problems expressed. For these, specially designed and executed efforts would have to be undertaken. It was particularly for this class of need that the program hoped to generate successful experimentation on the basis of interdepartmental co-operation.[15]

Federally sponsored programs such as 'Opportunities for Youth' and

'Local Initiatives Program,' under way at this time, were very well fitted to the efforts the community committees were making to discover their own needs and develop priorities for dealing with them. Thus the first class of needs, those the communities were able to act upon immediately, could be taken care of with these special funds. Unwittingly the federal initiatives provided just the kind of opportunities that RAP could prepare communities to take advantage of with local skills.

THE CONCRETE RESULTS

1 / A sound information base has been generated and published providing needed information for any efforts directed to planning development in the social, activity, and environmental aspects of southern Manitoba. This data base spans a period of twenty years so that long-term trends can be studied.
2 / The vast amount of Canada Land Inventory information about the environment has been transposed into usable form by an innovative system of composite natural environment capability maps which will permit detailed comparison of present land uses with the highly varied and complex existing natural environment conditions.
3 / A successful round of information exchange took place demonstrating that rural residents grouped into community committees for action, regional development corporations, government and private agency staffs, university based staff and students can effectively work together to raise the level of consciousness about living and working conditions in southern Manitoba.
4 / A highly valuable presentation in map plate format, accompanied by Working Paper no 2 on the 'Functional Relationships of Settlements and Their Spheres of Influence,' was produced as the synthesis response to the original request of RAP by government that some new understanding of community interrelationships be produced. Not only the content but the details of visual presentation of this information is a very helpful instrument to have at hand when trying to gain some insight into the community interaction network of southern Manitoba for future planning and development, whether at the scale of the whole region or in any part of the region.
5 / RAP played a key role in bringing about the reorganization of data aggregation by the line departments and agencies of government into 'Official Regions of Data Collection.' This is an extremely important achievement if up-to-date and complete information is to be available for

action programs. Many man-hours were spent in vain trying to reorganize information because of the previous hopeless tangle of records, data gathering, and storing.

6 / The native peoples of Manitoba were literally 'put on the map' in the RAP for the first time. Until this was done these people remained outside the realm of possible active participation in planning development projects.

7 / All the RAP map plates were produced in such a way that all the population of southern Manitoba formed a background with which to read any information explicitly shown on the plate. This technique along with the method of showing composite capability for the natural environment gives a dynamic quality to pertinent social and ecological information.

8 / Toll charges for telephone calls between users within designated areas have been removed so that now it is possible to have low-cost communication between dwellers in the interrelated settlements and farm areas of the province. Through the RAP studies it was discovered how penalized these people of low income were in having to pay long-distance charges for calls to adjoining settlements. Whereas a telephone user would have access to thousands of lines in Winnipeg, the rural residents had only a few hundred or less available on the regular charge basis.

9 / No record was kept of the spin-offs from the RAP exercise which occurred through regular government department activity. The relative gains in social and economic benefits can only be stated generally as having taken place.

ASSESSMENT OF RESULTS

In retrospect the results of RAP to date seem encouraging and enlightening. Encouraging – when one realizes all the pitfalls that could have interfered with the progress made in the organization and presentation of data; remarkable co-operation was experienced which allowed much to be done at very reasonable cost. Enlightening – because it became clear over five years that the 'system of rewards' within the civil service (as within a university system) is critical in attempting to introduce innovative responses to needs of people in rural Manitoba. Unless government staff are rewarded for horizontal or interdepartmental work as they are for intradepartmental work, one cannot hope to overcome the resistance of conventional practice in delivery of services.

RAP has demonstrated that it is possible to have an effective relationship between university, government, and communities when attempting to carry out social change in the rural areas of southern Manitoba. Despite the awkward timing of end-of-summer activities and the return to fall classes the over-all benefits of the arrangement were positive. However, this arrangement does demand that the government department and the university somehow modify the conventional organization of each institution so that continuity can be better achieved and a higher efficiency of output attained.

Perhaps the most telling insight of the experience thus far was seeing the extremely rapid build-up of complexity in trying to design, organize, and execute innovative responses to need on a 'cluster' basis. Prescinding from the political aspect of the situation, we can see how conflict can arise from a number of legitimate concerns. And this brings us back to the reward system – unless competition for internal power and budget control can be converted to some other source of job satisfaction, it is hard to see how seemingly obvious and simple projects can be successfully carried off, no matter how clear it is that they are greatly needed by the residents.

Within the RAP group it is clear that both creative thinking and leadership have to be combined with ability to work effectively with persons from a wide range of departments as well as with private agencies, regional development corporations, and rural residents. Reasonably successful efforts have been made in obtaining information from line departments and public and private agencies and thereby RAP was able to make considerable financial savings. The location of the RAP group within the government system is still a puzzle. As part of a line department it is inhibited in horizontal movement; as part of the Planning and Priorities Committee of cabinet secretariat it suffers an identity crisis.

In retrospect a critical event in the RAP process was the change in leadership of the group in 1973. Basically it was lost leadership as such as well as lost experience and growth in understanding about the innovative aspects of RAP that was involved. Those civil servants lost to RAP had helped in the sixties to develop some kind of regionalism in southern Manitoba, efforts which grew into six Regional Development Corporations for the area. It was these RDCs which in turn had greatly facilitated the initiation of RAP, the administering of a survey to five thousand rural families, organizing the December 1971 training seminar, moving the caravans through the rural areas of the province in the winter of 1972, and establishing the community committees for information exchange and reporting to government on their needs. At the time of writing, with the updated RAP information based on the 1971 census, there is available a vast array of information which is indispensable to any provincial government attempting to improve the quality of life for its citizens, especially for rectifying some of the disparity between rural and urban citizens.

With a somewhat biased judgment we think that RAP is a useful kind of instrument for identifying needed social and political change, for building a knowledge base essential for expressing effective concern for ecosystem conditions. It attained in practice a style of planning which was being written about by John Friedmann.[16] It grew on a solid base of experience tested at smaller scale. When one considers that at one stage over four thousand people were involved in the participation and information exchange process, the effort was worthwhile, for a 'learning process' has inevitably taken place, one which strengthens the base for future projects and future dreams for southern Manitoba.

CONCLUSIONS

What insights have been gained from the Regional Analysis Program in southern Manitoba?
1 / A style of planning that emphasizes learning as a part of the process can easily overlook the need for careful identification of the 'interests' of those who are the actors in this learning activity which accompanies (and is inherent in) regional planning. Our style of government has generally assumed that the legislators know better and more than the ordinary citizen. With the advance of all kinds of specialized knowledge this is a questionable assumption. By assuming that the recipients of any legislative program have a real contribution to make by way of essential information, there is a threat to legislators and civil servants involved. To say that the farmer and resident of small towns in rural Manitoba is an 'expert' in his own affairs, and about the environment in which he lives, clearly tells the departmental official that he must consult these experts as well as the group of technical people or scientists who are ordinarily dubbed 'experts.'

Identification of 'interests' then, ought to be seen not just at the decision level of activity in order to engage support or rejection of various issues; 'interests' ought also to be seen as information sources to be involved from the outset in any planning program. In RAP, the citizens of rural Manitoba were seen as experts from the beginning.

2 / Involvement of people in the gathering of information, in discussion about the make-up of their environment, about the characteristics of their town, and about the differences between their areas and others in the province, can quickly engender intelligent, determined, and forceful demands on government for ameliorative programs and projects. The people attain a new level of critical capacity to evaluate what the government is planning in the way of legislation. In fact, very quickly a ground swell of pressure can appear which puts the decision-makers in an awkward position; it demands of them some kind of planning strategy that can respond to a new seriousness and range of demands on the public system.

It seems to be quite clear that it was really the prospect of an election which finally led some cabinet ministers to take up in public discussion in 1973 the items of need raised in some of the community committee reports. It reflects the advantage of engaging commitment by government to a process through a transactive approach.

3 / The way in which information was gathered, organized, and exchanged, the degree of involvement by people through the work of the community committees, the clearly expressed desires by these people for a greater share in decision-making on matters affecting their local needs and aspirations (or at least wide consultation), clearly places a demand for an integrated planning process on the provincial government. No longer can easy trade-offs be made only on the basis of political debate in which persuasion rather than a thought-out scheme of planning is dominant.

4 / The strategy of holding off on setting any explicit goals for RAP, except for the two general ones of improving the quality of life and a concern for the impacts on the ecosystem of any changes carried out, allowed the real goals to be identified more clearly and be understood in the context of factual information. To say that the 'quality of life for rural Manitobans' was a basic goal was true enough; but until this was clarified further, the goals of 'stay option,'[17] 'equality of economic opportunity,' 'maintenance of environmental quality,' constructive interaction between what lay at hand and what was desired, could not get under way. Historical facts about settlements and the kind of natural environment they were parachuted into with the opening of the west in the last quarter of the nineteenth century can help us to appreciate the impact of today's decisions. We have seen how ecosystem inheritance can be squandered and the very life support system of the region greatly weakened, instead of being improved, by decades of land cultivation.

5 / In the realm of political activity, it is generally agreed that there has been increasing demand for participation in goal formulation, selection of means to achieve goals, and an effective sharing in the decision process affecting societal change. Even in rural areas this trend is under way. With growing awareness of the degree of interdependence between rural and urban functioning in any region, along with the centralizing nature of urbanization itself, more pressure is being brought to bear on politicians and planners alike to respond to the demands of the rural sector so that urban and rural differences might be lessened. At the same time the needed decentralization of certain operations of a large urban centre cannot be effectively achieved unless a cross-section of the population is willing to trade off the benefits of rural living for the sparkle (real or imagined) of the metropolis. In Manitoba, when effective decentralization of government operations took place, civil servants were reluctant to leave Winnipeg to live in smaller urban centres. Thus plans for decentralization were effectively frustrated. The civil servants cannot be blamed for reacting this way. What was not mentioned at the time was the key issue arising from the RAP experience – the unusual central-place hierarchy of Manitoba. There are no urban centres in the 50–75,000 class (roughly a tenth the size of Winnipeg). Except for Brandon (approximately one-twentieth the size of Winnipeg) there is no place to move to in Manitoba from Winnipeg unless one wishes to go to a centre *one-sixtieth* the size of Winnipeg. At this scale there is a critical deficiency of amenities for enriched human living. Compared to the metropolis, too much has to be sacrificed by people faced with a possible move to these small settlements. While the RAP experience brought out the fact that small centres function interdependently in groups, and a certain simulation of larger benefits can be provided by the group, there remains the crucial need to seek deliberately to establish urban centres of larger scale in order to offset the dominance of Winnipeg as a place to live. These are difficult political and environmental issues: political, because they call for well-informed and courageous decision-taking; environmental, because they call for an examination of the people-carrying capacity of southern Manitoba.

NOTES

1 Isaiah Bowman, *The Pioneer Fringe* (New York 1961), chap. 9
2 At this time the provincial government was about to *enforce* the amalgamation of the municipalities of the greater Winnipeg area into Unicity after ten years of metropolitan government.

3 'Regional Analysis Program' was the title of the program for the first year or so until it was changed to 'Rural Region Working Group.' Mention of 'regional planning' or of 'regionalization' was carefully excluded because at the time unease was widespread about rural regionalization in southern Manitoba.

4 This system of survey was imported from the United States, after much debate on technical details, following the transfer of the lands of the Hudson's Bay Company to Canada. Accounts of these complicated events in the opening of the Canadian west are provided in Chester Martin, *Foundations of Canadian Nationhood* (Toronto 1955), 407–92, and Don W. Thomson, *Men and Meridians*, II (Ottawa 1967), chap. 3.

5 'Urban' was defined by the Regional Analysis Program to mean settlements of 500 and more.

6 This denudation of the rural areas of population and the increase of population in urban areas is not unique to Manitoba; throughout Canada, the United States, and other parts of the world this phenomenon raises the question about the relationship between modern industrialization and the kind of urbanization that is going on. See the treatise by E.F. Schumacher, *Small Is Beautiful* (New York 1973), for a discussion of this critical problem. The UN Habitat meeting in Vancouver in 1976 brought out sharply how universal the problem of rural depopulation is, wherever modern technology has taken root.

7 Geddes developed the thinking of Frederic Le Play for practical use in regional planning. A society (folk) would only be understood when its occupation (work) and environment (place) had been thoroughly researched – cf. P. Kitchen, *A Most Unsettling Person* (New York 1975), 57–8.

8 There was no intention of producing a new atlas for the province in these map overlays. They were intended to focus very sharply some of the key issues that needed to be related to one another and a map format seemed the most suitable way of doing this. For example, the maps show both the dominance of Winnipeg over rural Manitoba and the lack of inter-community linkages by rail and by highway. At the very outset of any plan of action to bring about societal change the make-up of this inter-community linkage system must be examined in detail. When the west was opened towards the end of the last century it was precisely this system that was worked out so carefully in terms of the best available technology. Bands of track steel and telegraph wire linked grain-gathering points with distant shipping and processing places. Permanence for the system was seen in terms of efficiently moving grain out of the prairies and moving manufactures in. It was the mercantilist model transferred to Canada. And as we witness the erosion and fertility decline of prairie soils to maintain this outflow of grain we can more easily appreciate the close relationship between political economy and ecology.

9 In 1975 two volumes of updated statistics were published, based on the results of the 1971 census as well as information up to 1972 from provincial sources. By following the basic format of the original statistical output there are now available basic data for a twenty-year period.

10 Manitoba, *Guidelines for the Seventies* (Winnipeg 1973), I, 4, 10

11 The under thirty age group was chosen because the older generation were tiring from long years of struggle on the land, and frustration with the political process. A younger generation, which wanted to stay in rural Manitoba ('stay option'), needed an opportunity to demonstrate this interest in the future and take some part in bringing it about. In fact, as events unfolded there were many instances of the older folk being greatly

impressed by both the competence and commitment of the younger generation. It was, after all, because of concern for improving the opportunities for youth in rural Manitoba that RAP had been undertaken.

12 The design of the information exchange material was based on this use for which it was intended. The map folio, which at present has reached a total of 45 plates, was sized to fit the desks of the members of the legislature as well as to be a convenient size for rural working groups to carry about.

13 At mid-summer 1973 a provincial election was held and the New Democratic party was re-elected with a somewhat smaller majority.

14 Details of these solution packages were presented in a special consultants report by the program advisers in June 1972.

15 Details of these recommendations were contained in the 'consultants report' which did not receive public circulation and thereby become part of the information process.

16 Friedmann, *Retracking America: A Theory of Transactive Planning* (New York 1973)

17 'Stay option' is the phrase used to indicate the 'implementation of policies and programs which will prevent Manitobans from being coerced by economic forces to leave their province or to leave the region within the province in which they prefer to live.' *Guidelines for the Seventies*, I, 13

11
Environmental impact assessment:
reform or rhetoric?

Reg Lang

The ecology fad has long since come and gone. When our campus bookstore dismantled its ecology display (replacing it with science fiction and the occult), the demise of the fad was clearly at hand. Other outcomes of the environmental movement, however, have been more enduring: new environmental agencies to focus institutional action; sharply increased emphasis on pollution control, the re-use of 'waste' materials, and the protection of environmental resources; emergence of environmental law and environmental education as fields of study and practice; and a steady public concern which, though channelled into subsequent 'crises' such as energy and inflation, has been sustained at a level at least sufficient to keep politicians and businessmen wary of committing overt environmental blunders. As David L. Sills said: 'In countless other ways, the movement has influenced the major institutions of society and altered the behaviour of many people, whether they realize it or not, whether they like it or not.'[1]

'The environmental crisis' is typical of society's tendency to view as a single problem what in fact is a complex interplay of numerous related problems.[2] The corresponding tendency is to look for 'the big solution.' Environmental impact assessment has been so categorized, which makes it worthwhile to consider: what exactly is EIA, and to what extent does it offer real potential for effectively addressing environmental and social concerns, especially in Canada?

Environmental impact assessment forces explicit consideration of the consequences of proposed public and private interventions into natural and human environments. EIA can be defined as the systemic description, prediction, evaluation, and integrated presentation of the environmental effects of a proposed action at a stage where serious environmental damage may be avoided or minimized.[3]

EIA in this form was introduced in January 1970 when the United States National Environmental Policy Act came into effect.[4] NEPA

responded directly to environmental concerns urgently felt and expressed at that time by a sizeable body of public opinion. Public concern
focused on three interrelated aspects of environmental quality:

1 / *Activities*: Concern for man's ability to alter drastically the environment through population growth, urbanization, and large-scale massive
application of technology; and concern over where present trends were
headed.

Questioning of perceived narrow-interest, after-the-fact approaches to
environmental quality.

2 / *Environments*: Recognition of interrelatedness of the activity/en-
vironmental systems (web-of-life concept) and the bewildering com-
plexity/uncertainty this involved.

Recognition of limits to environmental capacities including resiliency
limits (within which systems can recover from stress) and thresholds of
capacity to assimilate human wastes.

Recognition of the importance of diversity of species and environments.

3 / *Effects*: Concern for environmental deterioration, pollution, loss of
valued resources, etc., which were perceived to affect a wide segment of
the population.

Concern for externalities and, subsequently, for distributive effects
(who benefits/pays?).

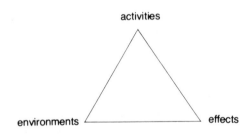

A specific and complete (by my definition) environmental impact
assessment would provide the following information:

1 / *The activity*: a full description and analysis of the proposal in question.

2 / *The environments*: a thorough inventory of the natural (ecological)
and human (socio-cultural) environments as well as interests affected by
the proposed activity.

3 / *Predicted environmental effects*: including (*a*) direct primary effects

together with indirect, secondary and tertiary (ripple) effects, (*b*) cumulative as well as initial effects, (*c*) beyond the site, effects on its immediate environs and larger region, (*d*) beyond individual effects to total effect, and (*e*) isolating of irreversible effects and irretrievable commitments. These effects are then linked to those who take the action, those who benefit from it, and those who bear the costs. Prediction of environmental effects involves comparing (*a*) the present condition *without* the project to (*b*) the future environment *with* the project. The concern is with how the environment might *change*, and more specifically, with change beyond that which would occur in any case. Probability and risk enter into this calculation. Timing and spatial dimensions of the predicted effects should also be included.

4 / *Evaluated environmental effects (that is, impacts)*: application of evaluation criteria, separating matters of 'magnitude' from matters of 'importance,' preferably through direct involvement of interests affected by the proposed activity (recognizing that while a degree of objectivity can go into predicting effects, differing and equally legitimate viewpoints will exist concerning what is important and to what extent);[5] and identification of distributive effects. Included here should be an explicit discussion of the evaluation criteria and how they were derived.

5 / *Mitigation*: description and environmental evaluation of possible mitigating actions that might be taken to offset adverse environmental effects.

6 / *Alternatives*: alternative means of carrying out the proposed activity (for example, a highway located here rather than there) and alternatives to the proposed activity (for example, public transit *vs* the automobile as a mode of transportation); and prediction and evaluation of the environmental effects of each alternative. The no-action alternative should be one of the options considered.

7 / *Presentation* of the results organized within an integrative framework that aids decision-makers in arriving at a final course of action (recognizing that the decision equation has two sides: one where information is gathered, knowledge is applied, as many variables as possible are identified, and understanding is gained; and the other where all this variety must be chopped down to a few actualities from which a choice can be made).

This is quite a formidable agenda, and one that is seldom followed, for various reasons: lack of time, money, and data (if something is impor-

tant, you can expect to have little time to accomplish it, which is the case
with many environmental impact assessments; unfortunately, some of
the problems researchers are trying to solve require long lead times to
amass the required data); lack of adequate knowledge and methods; lack
of a clear understanding of what the decision-makers require in order to
make a sound decision; and lack of that integrating framework which can
pull together the array of data produced by the foregoing seven headings
in a way that is comprehensible and facilitates decisions.[6]

It is not surprising, therefore, that the record of EIA is mixed, as the
following summary of US and Canadian experience indicates.

EIA IN THE UNITED STATES AND CANADA

Consider first the United States with nearly eight years of experience
under the National Environmental Policy Act. NEPA, which begins with
a dramatic declaration of national environmental policy,[7] has four main
provisions:
1 / *Explicit environmental goals for federal agencies*, for example, attaining the widest range of beneficial uses of the environment without
degradation, risk to health or safety, or other undesirable and unintended consequences; achieving a balance between population and resource use; maintaining an environment which supports diversity and
variety of individual choice; recycling depletable resources.
2 / *Environmental impact statements.* Section 102, the so-called
'action-forcing mechanism' and the guts of NEPA, directs all federal
agencies to 'include in every recommendation or report on proposals for
legislation and other major federal actions significantly affecting the
quality of the human environment, a detailed statement by the responsible official on (i) the environmental impact of the proposed action, (ii)
any adverse environmental effects which cannot be avoided should the
proposal be implemented, (iii) alternatives to the proposed action, (iv)
the relationship between local short-term use of man's environment and
the maintenance and enhancement of long-term productivity, and (v)
any irreversible and irretrievable commitments of resources which
would be involved in the proposed action should it be implemented.'

The environmental impact statement requirement provides for a
structured approach to environmental analysis of proposed actions, a
means for defining options, and a record of environmental considerations underlying a particular decision. More importantly, the EIS
constitutes a visible instrument, which, in combination with a sound

idea and a strongly felt concern, is necessary for any effective innovation.

3 / *Interagency consultation and public participation.* NEPA requires a systematic interdisciplinary approach which will ensure the integrated use of the natural and social sciences and the environmental design arts in planning and decision-making which may have an impact on man's environment, as well as the circulation of impact statements among agencies for comment, the development of appropriate new methods of analysis, and public dissemination of the results.

4 / *Creation of the Council on Environmental Quality* (CEQ) to oversee the EIA process, to issue guidelines under the act, to review and co-ordinate EISS, to review and appraise various agency programs to determine their environmental quality, and to prepare an annual report.

NEPA was an attempt to change the way federal agencies went about their environmental business, including how they planned for and made decisions and what considerations entered these processes. Note that NEPA is a full-disclosure law; that is, agencies are required to disclose and make available to the public, through impact statements, the full range of environmental effects of designated actions as well as other agencies' comments on them. But decisions concerning what action to take in light of the environmental assessment, and final decisions on whether or not to proceed with a particular action, remain the responsibility of the initiating agency. EIA seeks to raise the profile of environmental considerations in public decision-making but it does not give them a veto over other concerns. The political process of trade-off and decision remains intact.

From 1970 to 1976, nearly seven thousand draft environmental impact statements were filed by seventy federal agencies.[8] Transportation-related projects generated the most impact statements (over 40 per cent of the total); watershed/flood control projects were second (23 per cent), while the remainder spread over a wide range of activity. Initially, federal agency response to NEPA ranged from half-hearted compliance to outright avoidance.[9] In part this was due to the major overhaul of their decision processes demanded by NEPA, something few officials grasped initially. Subsequent experience with the act, plus pressure from the CEQ and considerable prodding by environmental organizations through the courts, appear to have resulted in the EIA requirements settling down to become part of most federal operations. And the innovation has spread. As of 1977, twenty-six states had some form of environmental impact assessment in place,[10] some of them (for instance, California) more stringent than NEPA; and numerous municipalities (as of 1975, nearly a third of all the US municipalities had adopted EIA approaches).[11] Environmental assessment seems to have gained a solid foothold in the process of governing in the United States.

Not so in Canada. While the Canadian version of EIA borrows freely from its US counterparts, its institutional form is considerably different.[12] Environment Canada's 'Environmental Assessment and Review Process' (EARP), created by cabinet directive in December 1973, operates through administrative policy and procedure rather than through federal legislation. Federal departments and agencies are directed to:

a / Take environmental matters into account throughout the planning and implementation of projects, programs, and activities that are initiated by a department or agency or for which federal funds are solicited or for which federal property is required.

b / Undertake or procure an assessment of potential environmental effects before commitments or irreversible decisions are made for all projects which may have an adverse effect on the environment.

c / Submit the assessments made for all major projects that will have a significant effect on the environment to the Department of the Environment for review.

d / Incorporate the results of environmental assessments and reviews in the design, construction, implementation, and operation of projects, giving environmental problems the same degree of consideration as that given to economic, social, engineering, or other concerns.

e / Include in program forecasts and annual estimates the funds necessary to carry out the intent of this policy and program.[13]

This procedure, however, provides little indication of the substantive environmental concern of the federal government. The operation of EARP is the job of Environment Canada's Federal Environmental Assessment and Review Office working through its Environmental Assessment Panels and its Screening and Co-ordinating Committees. EARP's stated aims are:

1 / To leave the management of environmental assessment and review in the hands of the proponent in order to avoid delay and decision-making responsibility;

2 / To provide an arm's length system of review advice and expertise; and

3 / To inform the public and, where appropriate, to involve the public in decision-making.[14]

It has been nearly five years since EARP was established and results have been slow in coming. As of 1976, only one impact study (for the Lepreau Nuclear Generating Plant in New Brunswick) had been completed and a mere eleven others were in preparation or being contemplated.[15] After a comparable period of operation of NEPA over four thousand draft or final environmental impact statements had been filed with the CEQ.[16] But action on the EARP front has picked up in the last couple of years. The latest federal registry lists four projects completed and twenty-one environmental assessment panels now in operation across Canada on a wide range of proposed activities (see Table 1). EARP is bringing various federal projects under environmental and public scrutiny, even though countless others escape. And recently the process

TABLE 1
Federal environmental assessment and review process:
projects completed, underway, and proposed

Completed
Alaska Highway Gas Pipeline, Yukon, 1977
Point Lepreau Nuclear Power Station, NB, 1975
Wreck Cove Hydroelectric Power, NS, 1976
Eldorado Nuclear Ltd. Uranium Refinery, Ont., 1978

Underway and proposed
Alaska Highway Gas Pipeline, Yukon
Arctic Natural Gas Pilot Project
Banff National Park Highway Improvements, Alta.
Bay of Fundy Tidal Power, NS
Boundary Bay Aerodrome, BC
CN Telecommunications System, Alta. and NWT
Dempster Pipeline, NWT and Yukon
Eastern Arctic Offshore Drilling
Eldorado Nuclear Ltd. Uranium Refineries, Ont. and Sask.
Deepening of Fraser River Shipping Channel, BC
Gull Island Hydroelectric Generation, Nfld.
Hamilton Airport Expansion, Ont.
Labrador/Nfld. Electric Power Transmission and Tunnel
Lancaster Sound Offshore Drilling, NWT
Mackenzie Delta Gas Gathering System
Mackenzie River Dredging Program
Polar Gas Project, NWT to southern Canada
Roberts Bank Bulk Loading Facility, BC
Shakwak Project, Alaska Highway
Vancouver International Airport Expansion, BC
South Yukon Transportation Study

Source: Environment Canada, Federal Environment Assessment Review Office, *Federal Environmental Assessment and Review Process, Register of Panel Projects and Bulletin*, no 4, June 1978. Copies of the register and information on EARP are available from FEARO, Ottawa K1A 0H3.

recorded a first: a panel recommendation against a project – a uranium refinery proposed by Eldorado Nuclear Limited at Port Granby on the north shore of Lake Ontario – as a result of concern over loss of prime agricultural land, air pollution, and disposal of low-level radioactive waste. The minister concurred and the panel is now getting ready to assess three other sites.[17]

Environmental impact assessment has also caught hold at the provincial level. By 1976 each province had some form of EIA in place.[18] Most provinces have proceeded under existing legislation, with new policies and procedures. One province, Ontario, went the NEPA route by passing a separate act with defined legal procedures, duties and rights of applicants and other proponents, and public access. Ontario's Environmental Assessment Act received third reading by the legislature in mid-1975. It is being proclaimed in sections, first to apply its provisions to designated 'undertakings' by provincial agencies, then to bring it to municipal government, and finally to apply it to the private sector. Section 5 (3) states that an environmental assessment shall consist of:

a / a description of the purpose of the undertaking;

b / a description of and a statement of the rationale for (i) the undertaking, (ii) the alternative methods of carrying out the undertaking, and (iii) the alternatives to the undertaking;

c / a description of (i) the environment that will be affected or that might reasonably be expected to be affected, directly or indirectly, (ii) the effects that will be caused or that might reasonably be expected to be caused to the environment, and (iii) the actions necessary or that may reasonably be expected to be necessary to prevent, change, mitigate or remedy the effects upon or the effects that might reasonably be expected upon the environment by the undertaking, the alternative methods of carrying out the undertaking and the alternatives to the undertaking; and

d / an evaluation of the advantages and disadvantages to the environment of the undertaking, the alternative methods of carrying out the undertaking and the alternative to the undertaking.

The part of the act applicable to provincial ministries and agencies came into effect in October 1976. Its results have yet to be felt, although EIA within the provincial government has been underway for some time. Proclamation of the sections applying the act to municipalities, still awaiting the outcome of a provincial-municipal study, is well behind the original schedule. Some private-project environmental assessments have been done voluntarily but so far the private sector remains unaffected by the act.[19]

EIA at the municipal level in Canada barely exists. Of the four ways in which US municipalities became involved in EIA – federal and state assessments that involve local government consultation, federal and state cost-sharing programs requiring EIA as a condition of assistance, state legislation mandating local EIA, and municipal initiative – only the last category properly applies to Canadian municipalities at this time. Only a few municipalities – for example, the City of Winnipeg and the Regional Municipality of Waterloo – have so far introduced environmental impact assessment into their planning and control processes.[20]

Comparisons between Canada and the United States must proceed with caution. In Canada much of the jurisdiction over environmental matters rests with the provincial governments; in that respect, it might be fairer to compare NEPA with the Ontario Environmental Assessment Act once the latter has gained some experience. United States–Canada comparisons are further complicated by: (*a*) differences in the way the two federal governments operate internally, for example, the relatively greater control the Canadian cabinet, compared with the executive branch or Congress, exercises over programs of federal departments; (*b*) differences in the role of the courts which play a policy-making role in the United States not assumed by their Canadian counterparts; (*c*) the relatively greater degree of secrecy and government restriction on 'public' information in Canada; (*d*) more constraints placed by the provinces on Canadian municipal governments compared with US municipalities which, in some states, enjoy considerable autonomy; and, finally, (*e*) the presence of a strong public demand for environmental action when NEPA was introduced, and the absence of such concern in Canada at that time.

Environmental impact assessment has arrived in Canada; of that there appears little doubt. But Canadian governments appear disinclined to opt for separate environmental assessment legislation, preferring instead to introduce it under existing acts, and by administrative arrangement. The Canadian approach is far less visible and considerably less accessible to the public, either directly through governmental processes or through the courts. It seems that EIA has mutated, and, although its influence is being felt in Canada, its various forms are not yet clear.

This raises the question: what can we expect EIA, whatever its final form, to do for us?

EIA PERFORMANCE

The results of evaluation depend heavily on the evaluation criteria –

which in turn depend on the answer to the question: what is the problem for which EIA is a solution?

At its roots environmental impact assessment is plain common sense: giving consideration to the full range of consequences of an action you are contemplating. As Dale L. Keyes puts it, referring to land use decisions: 'The emotionalism which has accompanied the use of the term "impact evaluation" in environmental debates has led to the notion that the words represent an innovative idea in decision making. A closer examination reveals that the term is fundamental to the very process of making decisions. Few would disagree that most, if not all, decisions are based on their likely outcomes, or impact. No decision to approve a subdivision, grant a variance or amend a zoning plan is made in a vacuum. Each is based on some analysis of the impact of making and implementing that decision. What is suggested here is a more comprehensive impact analysis procedure applied systematically to land use decisions. Rather than introducing a new idea, we are suggesting the expansion of an old one ... That is not to say that the suggestion is not somewhat disturbing. The usual constraints of time, money and knowledge, compounded by an intriguing web of vested interests, hidden agendas and political pressure militate strongly against procedures which may increase costs, tax knowledge and abilities, or improve the visibility of public decision making.'[21]

Obviously, then, there is more to environmental impact assessment than common sense. Environmental factors pervade society. Responses to environmental concerns, therefore, could fundamentally affect the way our society functions and, more importantly, the distribution of interests within it. This helps explain why exposing environmental consequences, and especially distributional effects, is threatening, but it does not adequately explain why environmental impact assessment had not been done until recently.

Brian J.L. Berry argues that the 'primary cause of environmental disruption' is 'industrialization under conditions in which environmental resources are undervalued.'[22] The key question is: why are environmental resources undervalued? The usual response can be expressed as a problem statement: 'Environmental resources are undervalued because public and private planners and decision-makers possess an inadequate awareness of the environmental consequences of the actions they are proposing and taking. With adequate information they would be more likely to recognize the true value of environmental resources and respond more sensitively to them.' Every problem statement contains the seeds of its solution. From the foregoing one, which has overtones of the rational model of decision-making,[23] the solution is readily apparent: ensure that planners and decision-makers obtain and give full consideration to the right kind and amount of information sufficiently early in the decision process so that alternatives may be considered and adverse effects avoided. While it is generally (though not always) acknowledged that better information does not necessarily produce better decisions, it none the less 'increases the probability that planners will have considered matters beyond the immediate economic and technical aspects of the proposed project.'[24] And so environmental impact assessment was born, and thrives.

NEPA appears to have been reasonably effective in forcing consideration of the environmental consequences of federal actions prior to commitment to them. Initially, agency response was slow, and many early impact statements were weak rationalizations of already committed action. Frederick R. Anderson, in the first major examination of NEPA, three years after its inception, noted: 'NEPA litigation has been primarily successful in stimulating after-the-fact rationalizations which are examined less by agency decision-makers than by agency lawyers whose job it is to ensure that the agency's environmental review can survive legal challenge.'[25] More recent experience appears more positive, however, leading the Council on Environmental Quality to report: 'With rare exception federal agencies with major EIS responsibilities reported that the EIS process is an important aid to planning and decision-making at various agency levels.'[26] And an in-depth study of three cases – an ocean dumping proposal, a sewerage system, and a winter sports complex – demonstrated that NEPA had real and worthwhile effects on the agencies' administrative processes even though the authors acknowledge that the complex mix of variables affecting governmental operations make generalizations extremely difficult.[27] Meanwhile, the CEQ (shaken by President Carter's initial attempt to cut back its operations, since changed to support) has undertaken a substantial inter-agency and public review of NEPA.[28] The result is a whole new set of regulations intended to strengthen the EIA process, ensure that environmental impact assessment is applied early in planning and decision processes, and reduce paper work and delays.

Finally, it appears that the mere threat of having to do an environmental assessment that is certain eventually to reach the public eye and possibly the courts is often sufficient to stimulate public agencies into a more environmentally sound form of management. It is also evident that

NEPA and its successors in the United States have opened significant opportunities for public involvement in government decision-making,[29] and have stimulated much-needed environmental research.

What the foregoing evaluation demonstrates are the *procedural* gains by NEPA: improvements in the way agencies plan, make decisions, and relate to their publics. And major procedural problems remain to be overcome; for example, 'the key problem for agencies has become how to measure and predict the significance of a physical or social impact.'[30] But what about the *substantive* aspects of 'the environmental problem'? Have environmentally improved procedures actually led to environmentally improved decisions and actions? Here the evidence is less clear, partly because it is often hard to say whether a project was halted or changed because of NEPA. The recent CEQ evaluation does provide a lengthy list of federal actions which it claims were stopped or significantly altered as a result of the EIA requirement.[31] But much remains hidden between the lines; objective evaluations have yet to be done.

Canadian experience can point to few such examples as yet. This country's involvement with environmental impact assessment is considerably more limited than that of the United States. But that alone does not explain the low profile taken by its governments in this regard and the almost negligible impact so far made on federal operations by environmental impact assessment. For example, the federal Environmental Assessment and Review Process has been criticized by A.R. Lucas and S.K. McCallum on three counts:

1 / EARP is largely internal and out of the public eye; public information and involvement with respect to environmental impact assessment are carefully controlled – and therefore minimal (to some extent this criticism has been overcome by recent public participation and hearing processes). Canada lacks the relative openness of public information enjoyed in the United States, even in those states without 'sunshine' laws.

2 / The proponent has considerable autonomy in determining whether his projects are to submit to EARP and, if so, what is to be assessed, how, and what attention to pay to panel recommendations (all of which ignores the argument that EIA came about precisely *because* proponent agencies were reluctant to consider environmental factors in their decisions). No legal avenue exists by which to force proponents to carry out environmental assessments. Canada's legal system does not provide citizens with access to the courts to the extent common in the United

States; and our judges do not play a legislative role as their American counterparts do.

3 / EARP has no independent watchdog comparable to the CEQ. And there appears to be no systematic attempt to monitor either the process of environmental impact assessment or the environments affected by actions that have been subjected to EIA.[32]

It is questionable how much priority the federal government gives to EIA (and environmental issues generally), and even why it allows environmental impact assessment to exist. The cabinet seems to have concluded that public concern for environmental degradation does not match the political risks; on the other hand, there is the attraction of internalizing, thereby controlling, the EIA process. In the United States the Council on Environmental Quality noted in its third annual report that 'citizen enforcement of NEPA through court action has been one of the main forces in making the Act's intended reform a reality.'[33] Canadian governments, ever alert to learning from US experience but fearful of runaway innovations, do not appear willing to risk the uncertain outcome of court action on environmental projects – especially since some of Canada's most controversial decisions (James Bay, Mackenzie Valley, etc.) can easily be seen to have significant environmental consequences.[34]

Ontario's legislation seems to promise new opportunities for environmental impact assessment. The act is dedicated to 'the betterment of the people of the whole or any part of Ontario by providing for the protection, conservation and wise management in Ontario of the environment'; and the minister who introduced the act called it 'one of the most important pieces of legislation ever introduced in Ontario.' But the provincial government's real commitment to EIA is seriously in doubt. A recent newspaper editorial observed: 'And this month, as we celebrate the third anniversary of the passage of the Environmental Assessment Act, there is no alternative to the conclusion that the act has withered on the vine, been subverted from the outset, is a sham, a subterfuge, a bust. All show, no go.'[35] So many projects have been allowed to bypass the legislation that it has come to be known as the Environmental Exemption Act. The Ministry of the Environment has received only five project assessments (four are still under review), compared with the hundreds that have been exempted, and no public hearings have been held under the act by the environmental assessment board. Other provinces (notably Alberta) which chose to pursue environmental impact assessment with less fanfare and fewer heroics appear to have accomplished more.

Where does all this leave us? Environmental impact assessment,
according to the us experience, has considerable potential as *a* solution
to the problem of how to force government agencies to consider environmental factors in their planning and decision-making. This potential seems to have a much greater chance of being realized when:
a / EIA has a solid legislative base, though not necessarily in a separate
act.
b / The legistative/policy/administrative provisions clearly bring and
keep EIA out in the open, spelling out the process to be followed
alongside the rights of the public to know about, request, and challenge
environmental assessments; and thereby providing grounds for citizen
action in the courts.
c / Information concerning activities requiring environmental assessment is accessible to the public.[36]
d / A watchdog agency oversees the EIA process, ensuring governmental compliance to its provisions and regularly monitoring and an-
nually reporting on the outcomes.
    Even if all these conditions prevailed, problems would remain. One is
Canada's scarcity of large-membership environmental organizations
with sufficient capability and clout to intervene effectively on behalf of
environmental concerns; consequently, the lack of access to financial
and technical resources (most of which are commanded or strongly
influenced by governments) constitutes a barrier to an effective environmental assessment process.[37] A deeper problem is that environmental impact assessment, even as rigorously applied under NEPA, does
not deal with most of the actions which, singly and in combinations and
over time, produce environmental degradation. Four observations can
be made here.
    First, many us federal actions escape NEPA: 'the EIS has received little
emphasis as a major departmental policy aid in the Department of
Defense, with the exception of the Corps of Engineers.'[38] Similarly, in
Ontario a whole range of ministries are exempt from the Environmental
Assessment Act; this includes Housing, Agriculture, Education, and
Health. The actions of other governmental levels can be expected to
contain similar glaring omissions.
    Second, environmental concern has not penetrated significantly into
the actions of the private sector, as a recent detailed study concluded:
'twenty-one recently-planned projects, ranging in cost from 35 million to
5.5 thousand million U.S. dollars, were investigated by means of inter-
views with executives of seventeen petroleum, chemical and metal firms

in nine different nations. A broad range of planning behaviour was discovered. At one end of this range were a few firms which appeared to be comprehensively taking ecological considerations into account. At the other end were many more which appeared to be indifferent to ecological concerns and were considering little beyond the immediate economic and technical aspects of their projects. The central conclusion of the research was that the practice of ecologically-oriented industrial project planning is still in its infancy.'[39]

Third, large-scale actions tend to escape the full effects of environmental impact assessment; a prime example is Ontario's huge Darlington nuclear power plant, forecasted to cost $3.5 billion! These are precisely the actions which carry the greatest environmental risk. They also have long lead times and tend to be committed well before they become public knowledge. And they are the most politically hazardous, dealing with long-term futures, basic human needs, fundamental conflicts, and issues of who gets what.

Finally, environments are degraded less by discrete actions than by the combined and cumulative effects of many actions in complex and often incalculable ways that defy advance assessment. For example, it may not be a trunk sewer or a road that significantly changes an environment but rather the subsequent pattern of urban 'development' attracted by excess service capacity. Environmental impact assessment processes, therefore, must not only predict and seek ways to mitigate adverse environmental effects but also monitor and evaluate the actual effects irrespective of their causes and initiate corrective responses. Pre-action and post-action evaluation are unquestionably integral. Yet EIA legislation focuses only on the former, ignoring or giving mere lip-service to the latter.

BEYOND ENVIRONMENTAL ASSESSMENT

Environmental concerns obviously go much deeper than environmental impact assessment is able to reach. Sills observes: 'Environmental policies both cope with and increase scarcities of various kinds; they encourage a certain type of life-style and discourage other types; they activate interest groups; they create conflicts over environmental decisions; and they both influence and are influenced by those who hold power in society. In short, they are political policies, and since the environmental movement seeks to influence the formulation and execution of these policies, it is in the broad sense of the term a political

movement.'[40] The most telling criticism of environmental impact assessment (and the most unfair, since no one measure could achieve this)
is that it fails to get at the roots of 'the environmental problem,' at the
basic forces (high-consumption lifestyle, powerful interests deeply
vested in activities detrimental to environmental quality, etc.) that generate environmental degradation and the need for EIA in the first place.

Some would argue that EIA, as a product of our (insert your favourite
adjective) system, inevitably must support that system. Probably so, in
the broadest over-all sense. Such critics usually are looking for un-
specified but none the less dramatic large-scale change. Even if it could
be argued convincingly that a sudden overthrow of our government and
other key institutions was possible and desirable, what would be the
likelihood of their replacements being more environmentally oriented?
Unlikely, if countries with alternate approaches to government are any
indication. Socialist nations, whether democratic or not, do not seem to
have a better environmental track record than we do.[41] Nor do countries
unburdened by the Judeo-Christian tradition with its system of allowable beliefs and behaviours that condone man's conquest of nature. The
North American environmental problem, instead, seems to be a result of
a number of factors in combination: 'America is thus the archetype of
what happens when democracy, technology, urbanization, capitalistic
mission, and antagonism (or apathy) toward natural environment are
blended.'[42]

'The problem' is more likely a syndrome – a group of factors acting
together, perhaps with synergism – that builds up incrementally. 'The
solution,' therefore, must possess similar characteristics, a clustering of
measures applied incrementally, but with a *common purpose*.[43]

Clearly, this is a political process. But it cannot just be the politics we
have now. Environmentally, it would be self-defeating to rely on current
political processes which undervalue environmental factors, as they do
the well-being of the less privileged in society. The two are closely
related at this point, for a characteristic of that North American blend
described above is that it distributes its costs and benefits quite unevenly
among its members. And so the poor suffer most from deteriorated
environments – living in the most unsatisfactory housing, in inner cities
where the air is heavily polluted and the noise is most oppressive, or in
rural slums with the fewest available means for weekend escape to better
environments. They are also likely to pay a disproportionate share of the
costs to remedy environmental problems. 'As a result of these influences
– a slowing down of economic growth, price increases, regressive taxa-

tion, opportunity costs – environmental improvements are said by critics to be regressive in their distributive effects,' Sills concludes.[44]

The central core of the common purpose referred to above must be environmental quality *and* social equity. Each pursued alone, whether separately or within a common concept but in tandem, is likely to be at the expense of the other. Pursuing them together will be no easy task, though. What we were unable to do in times of plenty will be even more difficult in times of scarcity; on the other hand, redistribution can be more easily avoided under the former conditions – an ever-bigger pie implies bigger slices for all – and perhaps more easily disguised under the latter condition. Still, there is the fundamental barrier experienced by 'a society that possesses the technological means to afford abundance, but cannot display those means to meet the needs of its poorer members because this would be politically intolerable to the marginally affluent, work-oriented mass of its constituency.'[45] One cannot help but feel that much of Canada today faces an economic future in which the marginally affluent group, up against the wall financially, will grow dramatically, not because more people below that margin are improving their incomes but because more people above it find themselves increasingly less well-off. At the same time, as this group becomes larger and as more of its members perceive fewer chances of escaping (there can never be enough lotteries), inequity may become a hot political issue. And if more and more people crowd into high-density living in a few large cities, attracted there by perceived economic opportunity but unable to afford the commuter's or the cottager's escape, we can expect the distinction between environmental inequity and social inequity to blur until the two seem as one.

Under those conditions environmental impact assessment could take on a different and far more significant role than it has at present: a dynamic instrument for social change. Even in its present form, EIA has the potential to move towards that role, in two respects.

First, environmental assessments begun early in planning processes can address the need underlying a proposed activity ('what is the problem for which this is the solution?'). This is essential if we are to: (*a*) break certain high-consumption patterns (for example, in the fields of energy and waste disposal, based on erroneous notions of inexhaustible resources and unlimited assimilative capacity of natural systems); and (*b*) close the circle on people and institutions who are now able to ignore the wider and deeper consequences of their definitions of what is needed and how that need will be met.

Second, environmental assessments can be used to reveal present distributional inequities and forecast expected ones resulting from proposed actions. Such information is now either absent or a well-kept secret; its continued absence is a necessary condition for the maintenance of the inequitable pattern. Environmental impact assessment, especially if combined with post-action monitoring and evaluation, together with the establishment of community data bases for this purpose, could be a leading edge in establishing 'distributive impact' as a normal consideration in planning and decision-making.

Many barriers to this kind of thinking and acting obviously exist. One is the rational/objective/professionalized style of public management that has a cozy power relationship with the existing order and that thus resents outside interference, dreads exposure of what are here called consequences but may also be seen as errors or omissions, and has its own vision of what is best for the people. In these hands tools such as EIA take on a different form than originally intended. That should be no surprise, though; tools are value-free. EIA undoubtedly has been and will continue to be used to support as well as stop bad projects.

Environmental impact assessment: reform or rhetoric? The answer is yes, some of each, because they are related. The rhetoric of EIA stems in part from an astute political awareness of the risk it creates for unintended reform. As a measure aimed at forcing environmental considerations into decision-making EIA is an *intended minor* reform. What makes it different, and what raises the prospect of *unintended major* reform, is the environmental assessment *process* which legitimates, brings together, and provides a forum for examination of some sensitive issues, central to the way our society now operates, which otherwise tend to be kept under cover and apart. Need and the distribution of costs/benefits have been cited. In addition, people directly and adversely affected by a project get a say in key decisions (they are usually opposed) and their views are brought into contrast with those (usually farther away) who favour the project and experience a better cost-benefit ratio. The often tenuous basis for decisions (certain environmental standards, for example) is exposed and behind-the-scenes giving of scientific advice is forced out into the open for challenge. Matters of fact and matters of value are separated and both become relevant; blurring the distinction tends to favour the former and makes project decision-making the exclusive domain of technicians. Advisers and decision-makers alike are brought face-to-face, some for the first time, with the people whose lives and environments their actions affect directly. A fair number of people, politicized in the process, go away wondering not just what the hell is going on *here* but in general. And the publicity their case receives can be manipulated to link up with concerns of wider constituencies, perhaps creating a serious political issue.

Environmental impact assessment is risky and therefore must be carefully controlled. Risk generates rhetoric; rhetoric disarms, even substitutes for, reform. This strategy works as long as the rhetoric remains convincing, especially to those well-intentioned people inside the system whose strong desire for environmental improvement coupled with commitment to their organizations blinds them to the reality and makes them vulnerable to exploitation by the very forces they hope to change. Yet, the fact that environmental impact assessment produces such rhetoric signals its significant reform potential. The stronger the rhetoric, probably the greater the fear of and need for such reform.[46]

That should be enough to cause those of us who seek reform to give environmental impact assessment a close, careful look, and to nurture its reform potential. Too few such opportunities exist for us to treat any of them lightly.

NOTES

The key ideas presented in this paper evolved from discussions with Audrey Armour as part of our work to develop an environmentally oriented mode of planning/management that is congruent, both procedurally and substantively, with the requirements of environmental problems.

1 Sills, 'The Environmental Movement and Its Critics,' *Human Ecology*, III, 1 (1975), 6
2 Chevalier calls these metaproblems: large complex problems which have many facets but are perceived by society as a single problem; affect large numbers of groups and individuals with varied and conflicting interests; encompass multiple and conflicting ends and immeasurable connections between ends and means; may be so large and multi-faceted that they affect every part of society, yet no one major organization has the responsibility or vested authority for them; and must be dealt with without fully understanding them. Michel Chevalier, 'A Strategy of Interest-Based Planning,' Ph.D thesis, University of Pennsylvania, 1968. For another treatment of a similar subject, the 'wicked' problem, see Horst W.J. Rittel and Melvin M. Webber, 'Dilemmas in a General Theory of Planning,' *Policy Sciences*, 4 (1973), 155–69.
3 Environmental impact assessment, in concept and method, is closely related to technology assessment which, focusing more broadly on technological change rather than specific actions, has been defined as follows: 'Technology assessment is the process of taking a purposeful look at the consequences of technological change. It includes the primary cost/benefit balance of short term, localized, market place economics, but particularly goes beyond these to identify affected parties and unan-

ticipated impacts in as broad and long range fashion as possible. It is neutral and objective, seeking to enrich the information of management decisions. Both good and bad effects are investigated since a missed opportunity for benefit may be just as detrimental to society as an unexpected hazard.' R.A. Carpenter, cited in M. Gibbons and R. Voyer, *A Technology Assessment System: A Case Study of East Coast Offshore Petroleum Exploration* (Ottawa 1974), 24

4 The act itself, its guidelines, and a summary of current experience are contained in annual reports of the Council on Environmental Quality (CEQ) beginning with the first report in 1970.

5 The distinction between magnitude and importance is elaborated in L.B. Leopold *et al.*, *A Procedure for Evaluating Environmental Impact* (Washington 1971).

6 An example of an EIA that did *not* do this: US Department of Housing and Urban Development, *Draft Environmental Statement on the New Community of Cedar-Riverside, Minneapolis, Minnesota* (Washington 1974), a 233-page document that lists numerous effects but does not deal adequately with their incommensurability. In March 1976 the US District Court in Minneapolis ordered that the project not proceed on the grounds that the Environmental Impact Statement was inadequate (failure to discuss alternatives adequately; improper balancing of financial/technical considerations with the environmental; weak and vague discussion of long-range environmental impacts and resource commitments). See Rodney E. Engelen, 'Cedar-Riverside: A Case Study,' *Practicing Planner*, American Institute of Planners (April 1976), 30–40

7 The act's preamble states as its goal: 'to use all practicable means and measures, including financial and technical assistance, in a manner calculated to foster and promote the general welfare, to create and monitor conditions under which man and nature exist in productive harmony, and to fulfill the social, economic and other requirements of present and future generations of Americans.'

8 CEQ, *Environmental Impact Statements: An Analysis of Six Years' Experience by Seventy Federal Agencies* (Washington 1976)

9 CEQ, *Environmental Quality: The Fourth Annual Report of the Council on Environmental Quality* (Washington 1973), 244

10 CEQ, *Environmental Quality: The Eighth Annual Report of the Council on Environmental Quality* (Washington 1973), 130–5

11 For example, see Steve Carter *et al.*, *Environmental Management and Local Government* (Washington 1974). California has progressed the furthest in the application of EIA to local government; see Arthur W. Jokela, *Self-Regulation of Environmental Quality: Impact Analysis in California Local Government* (Claremont 1975)

12 A federal Task Force on Environmental Impact Policy and Procedure recommended, in August 1972, a NEPA-like approach for Canada. What became of this advice is unknown and the report remains confidential. A summary of the task force terms of reference and recommendations of its report are contained in Reg Lang and Audrey Armour, *Information Resources for Environmental Assessment* (Toronto 1978).

13 Environment Canada, Ontario Region, *Environmental Assessment in Ontario: The EARP Process* (Toronto n.d.). See also Fisheries and Environment Canada, Environmental Assessment Panel, *A Guide to the Federal Environmental Assessment and Review Process* (Ottawa, Feb. 1977)

14 A.R. Lucas and S.K. McCallum, 'Looking at Environmental Impact Assessment,' in P.S. Elder, ed., *Environmental Management and Public Participation* (Toronto 1975)

15 Reg Lang and Audrey Armour, *Urban Environmental Assessment in Canada and the United States* (Ottawa 1976), app. A.6, 'Federal EIA Approaches'
16 CEQ, *Environmental Impact Statements*
17 Environment Canada, *Report of the Environmental Assessment Panel on the Port Granby Uranium Refinery Proposal, Eldorado Nuclear Ltd.* (Ottawa 1978)
18 Lang and Armour, *Urban Environmental Assessment*, app. A.7, 'Canada: Provincial EIA Approaches'
19 Refer to Ontario, Ministry of the Environment, 'Environmental Assessment Required of Major Ontario Government Projects,' *News Release*, 19 Oct. 1976; and *Regulation Made under the Environmental Assessment Act, 1975* (21 Sept. 1976). See also the ministry's periodical, *EA Update*, especially II, no 1 (Jan. 1977), to which is appended the 33-page 'Report of the Municipal Working Group, Recommendations for Designation and Exemption of Municipal Projects Under the Environmental Assessment Act.' The Ontario setting and the act are described and evaluated in Mary Jane Carswell and John Swaigen, eds., *Environment on Trial: A Handbook of Ontario Environmental Law* (Toronto, rev. ed. 1978).
20 Audrey Armour and John Walker, 'Canadian Municipal Environmental Impact Assessment: Three Case Studies,' *Plan Canada*, 17, no 1 (March 1977), 28–37, which documents the Winnipeg and Regional Municipality of Waterloo examples at that time. Also see Sharon L. Earn, 'Environmental Assessment and Municipal Planning: Problems and Prospects,' *ibid.*, 38–47.
21 Keyes, *Land Development and the Natural Environment: Estimating Impacts* (Washington 1976), 2
22 Berry and Frank E. Horton, *Urban Environmental Management: Planning for Pollution Control* (Englewood cliffs, NJ 1974), 12
23 Gibbons and Voyer, *A Technology Assessment System*, summarize the rational approach which exhorts the decision-maker to (1) identify, scrutinize and order the objectives governing choice of a solution to the problem; (2) comprehensively survey all possible means of achieving these objectives; (3) exhaustively examine all possible consequences of each alternate means; and (4) choose the means that maximizes achievement of the desired objectives. See also Y. Dror, *Public Policymaking Reexamined* (San Francisco 1968), chap. 12
24 Thomas N. Gladwin and Michael G. Royston, 'An Environmentally-Oriented Mode of Industrial Project Planning,' *Environmental Conservation*, II, 3 (Autumn 1975), 196
25 Anderson, *NEPA in the Courts: A Legal Analysis of the National Environmental Policy Act* (Baltimore 1973), 287. See also Frank Kreith, 'Lack of Impact,' *Environment*, XVI, 1 (1973), which sampled EIS s and showed that actions proceeded essentially unchanged despite adverse comments and predictions of environmental degradation; and William V. Kennedy and Bruce B. Henslaw, 'The Effectiveness of Impact Statements: The U.S. National Environmental Policy Act of 1969,' *Ekistics*, 218 (Jan. 1974), 19–22, who examined 200 final impact statements filed in 1971 and found EIA guidelines full of ambiguities, lack of consistency in preparation of statements which ran from a few pages to volumes, many statements lacking coverage of social factors, difficulty for citizens to get hold of information in time to participate effectively, and lack of adequate federal funds to back up NEPA's administration.
26 CEQ, *Environmental Impact Statements*, 73
27 Jeffrey Shane *et al.*, *NEPA in Action: The Impact of the National Environmental Policy Act on Federal Decision-Making* (Washington, Oct. 1975)

28 See *Federal Register*, Part II, 9 June 1978. Also see Commission on Federal Paper-work, *Environmental Impact Statements* (Washington 1977)

29 Circular A-95 of the US Office of Management and Budget, issued in 1969 with sub-sequent amendment, established the 'Project Notification and Review System,' a form of early warning system to provide timely information on federal proposals and to assure co-ordination of stated plans, programs, and projects in other federal agencies and with state and local levels. A-95 established state and area-wide clearinghouses which – and this applies to EIA – are to identify jurisdictions and agencies (including state and local governments as well as non-governmental public organizations) whose interests might be affected by a proposed project and give them an opportunity to participate in its review. An example of the procedures used is given in Capital district Regional Planning Commission, *A-95 Project Notification and Review System: Procedures Guide for Applicants in the Capital district Region* (Albany 1976).

30 CEQ, *Environmental Impact Statements*, 50

31 CEQ, *ibid.*, app. D, 'Examples of the Effect of the EIS Process on Federal Decisions'

32 Lucas and McCallum, 'Looking at Environmental Impact Assessment,' 311–13.

33 CEQ, *Environmental Quality, The Third Annual Report of the Council on Environmental Quality* (Washington 1972), 248

34 In the case of the Mackenzie Valley pipeline, the federal cabinet chose to use a different approach, appointing a one-man board of inquiry, headed by Justice Thomas R. Berger, into the broad social, economic, and environmental impacts that a gas pipeline and energy corridor would have in the Mackenzie Valley and western Arctic. Berger's unique, human style – going out into the remotest native settlements and actually listening to people – and his controversial conclusions and recommendations would appear to have guaranteed a long waiting period before this approach will be tried again. See *Northern Frontier, Northern Homeland: The Report of the Mackenzie Valley Pipeline Inquiry*, I and II (Ottawa 1977 and 1978)

35 'Hollow As a Promise,' *Globe and Mail*, Toronto, 12 July 1978. Public hearings on environmental assessment have been held in Ontario (the current series on Elliot Lake uranium mining is an example) but under the Environmental Protection Act or orders-in-council rather than under the Environmental Assessment Act. The distinc-tion is significant. Under the latter act the environmental assessment board has wide scope regarding what can be considered, and it decides on both the adequacy of the environmental assessment and whether the project should proceed. In other hearings the board's activities are constrained and its role is merely advisory.

36 The need for 'freedom of information' legislation is becoming a significant political issue in Canada, with such a bill now before Parliament, legislation promised in Ontario, and legislative committees across the country instigating the means whereby government information may be made readily available to the public without jeopar-dizing those few key government activities that demand confidentiality.

37 Recently some environmental groups have received government funding to support their interventions in environmental assessment processes. Examples are Ontario's Royal Commission on Electric Power Planning and its Royal Commission on the Northern Environment; the Mackenzie Valley Pipeline Inquiry; and Saskatchewan's Cluff Lake Board of Inquiry. Unfortunately, these are rare exceptions to the general pattern. Probably the main barrier to funding of public interest groups is a simplistic but popular assumption: since governments already represent the people, it is unreason-able to use public money in support of efforts to oppose government actions.

38 CEQ, *Environmental Impact Statements*, 24
39 Gladwin and Royston, 'An Environmentally-Oriented Mode of Industrial Project Planning,' 189.
40 Sills, 'Environmental Movement,' 26
41 Sills, *ibid.*, 33, cites several sources to refute the thesis that capitalism inherently leads to pollution while socialism does not. See also Berry and Horton, *Urban Environmental Management* 11–12
42 Berry and Horton, *ibid.*, 11
43 Bertram Gross has argued, '... major structural change – even change of a revolutionary nature – cannot take place except through a series of small steps. These steps, however, must follow each other in accordance with some broad strategic considerations. They must be jointed, instead of disjointed. Hence, jointed incrementalism is one of the major principles of strategic decision-making.' Cited in Donald M. Michael, *On Learning to Plan – and Planning to Learn* (San Francisco 1973), 63–4
44 Sills, 'Environmental Movement,' 32, which includes an array of sources to back up this contention. See also James N. Smith, ed., *Environmental Quality and Social Justice in Urban America* (Washington 1974)
45 Edgar Z. Friedenberg, *The Disposal of Liberty and Other Industrial Wastes* (Garden City, NY 1976), 1
46 This seems to be the case in Ontario where the provincial government's glowing language describing environmental assessment co-exists with its increasingly dismal environmental record. In recent months the government: substantially reduced the protection area covered by the Niagara Escarpment Commission; settled a lawsuit, begun in 1970 against Dow Chemical for alleged mercury pollution of Lake Erie and Lake St Clair, for a small fraction of the \$35 million claimed; appointed an industrialist to head the Royal Commission on the Northern Environment; gave Reed Paper more time to clean up pollution it was told to stop eight years ago; and ignored evidence of acid rain effects and extended the order permitting Inco in Sudbury to continue discharging until 1982, 3600 tons a day of sulphur dioxide into the air, thereby backing away from an earlier order that would have reduced that level to 750 tons by 1978.

12
Political aspects of environmental issues

William Leiss

In his *Report of the Mackenzie Valley Pipeline Inquiry* Mr Justice Thomas Berger recommended that no energy corridor be permitted in the coastal region of the northern Yukon. He proposed the creation of a national park in that area – a proposal that has now been accepted by the federal government – which would 'afford absolute protection to wilderness and the environment by excluding all industrial activity within it.' The park will protect an area vital to the maintenance of the Porcupine caribou herd, snow geese and other birds, fish, and many other wildlife species. On the day the Berger report was released, the television networks sought responses from Canadians on its principal recommendations. Many persons erroneously interpreted the wilderness park proposal as a denial of Arctic natural gas to southern Canada. One of the responses came from a well-to-do Toronto suburban couple: Asked whether they would forgo natural gas for their home for the sake of the Yukon caribou, they replied, with barely concealed outrage, 'It's them or us.'

Echoing the tone of this response, in its mixture of indignation and bewilderment, are some of the most fascinating paradoxes of contemporary society. After centuries of industrial development and the 'conquest of nature,' after eliminating most of the wildlife and their habitat on the North American continent, we find that we cannot afford to become sentimental about some of the remaining caribou and geese who unwisely choose to breed near substantial fossil fuel deposits – regrettably so, for were we not in such dire need of the substances, we would be happy to consign that unpleasant territory to them and the few peculiar humans who apparently like living there. The needs of the industrial machine, like those needs it services, know no natural bounds.

The chief paradox is the pervasive insecurity of a society whose economy's material output is so abundant, as measured by most standards derived from past human history. The larger this output becomes the more carefully it must be watched, and governments today have little time for anything except nursing the Gross National Product. Our insecurity stems mainly from the curious impermanence of what we want and how we produce it. Very little in our system is self-renewing, save the wants that drive it on, and thus we must search in ever more remote places, with escalating costs and more esoteric technologies, for materials and energy to feed it.

The prevailing rules in our regulated market economy encourage a rapid turn-over of wants and products; industrial concentration and complex technologies make available for such purposes hitherto inaccessible deposits of the earth's resources. Prodigious waste of those resources seems to be a necessary function of this accelerating turn-over, this impermanence of wants and products. (Barry Commoner estimates that eighty-five percent of the capacity to do useful work in the energy we use is wasted.[1]) Conservation, recycling, and re-use would hold up the appearance of the 'new.' This profligacy has its price: will we have enough for the next round? Collective insecurity mounts in proportion to the growing heaps of half-used artefacts and momentary impulses flung into the surrounding disposal sites.

Thus the Porcupine caribou herd appears to threaten our well-being. The apparent threat involves the real issue of making Arctic gas and oil available for southern markets in Canada and the United States. It has symbolic overtones as well. For the response 'It's them or us' is only a fleeting expression of the more general collective insecurity that pervades our present political economy. And it is this insecurity that for the most part defines the politics of environmental issues today.

What are here referred to as 'environmental issues' are, in the most general terms, problems concerning (1) the impacts of different technologies (especially industrial technologies) on the regenerative capacity of natural ecosystems and on human health, and (2) the impacts – on both human societies and non-human species – of modifying habitats for settlements or other uses. The politics of environmental issues, considered likewise in the most general terms, concerns what we choose to do (or not to do) about those issues – that is, what degree of importance we attach to them, how well we understand the nature of the impacts, and what effect our attending to them has on our institutions and social relations.

In one sense everything we do has an environmental impact: each of us, at every moment, is involved in complex interactions involving the organic and inorganic chemistry of nature. Most of these 'impacts' cannot be considered, except in a trivial sense, to give rise to environmental issues. In this as in other cases, a deliberate focusing of interest and awareness is a necessary aspect of what we mean by referring to something as an 'issue.'[2] Thus issues normally arise with respect to social problems that are already well developed. On the other hand, statements of issues often miss the problem entirely, and there are also pseudo-issues that invent both the problem and its putative solution simultaneously.

When a set of issues first comes to public attention – as environmental issues did in the late 1960s – it can have a dual significance. It can be expressed as demands for solutions to immediate practical problems and also as a broad perspective on social development. For example, the response to nineteenth-century industrial working conditions encompassed both specific improvements in factory settings and – in Marxism – an all-embracing theory of social evolution. The prominence of environmental issues in the preceding decade has been expressed both in practical terms (such as better air-quality standards) and in more frequent discussions of long-range perspectives (ethology, sociobiology) in human ecology.

The reasons why a heightened sensitivity to particular issues occur at a certain point are difficult to determine. What is clear is that familiar events come to be seen as a 'pattern': we *construct* a new way of seeing the interrelationships among things. For example, sex-role stereotyping can be taken for granted, as if women had an innate aptitude for stenography or answering telephones, until these conditions are understood as part of a socialization pattern that permeates an entire culture. Simply reversing the familiar stereotype in this case, as was done in Norman Lear's soap-opera series, 'All That Glitters,' can enable us to see more clearly not merely the sex-role stereotyping, but many other features in our social relationships as well.

For many persons the heightened sensitivity to environmental issues also meant a new way of seeing connections among different concerns. Air and water pollution, waste of resources, occupational health hazards, and the extinction of wildlife species began to be seen as elements of a pattern of social action, a pattern that was built upon both institutional arrangements and deeply rooted collective attitudes. Familiar doctrines and events are cast in a new light and 'fall into place' in the new pattern. Lynn White's 1966 lecture, 'The Historical Roots of Our Ecologic Crisis,' which placed contemporary environmental concerns in the context of Western religion, created a minor sensation and has been reprinted in almost every collection on environmental issues since then.

One of the dangers involved is that the new way of seeing things can become the exclusive way: everything now appears as an illustration of this one pattern. It runs the risk of falling victim to the laws of fashion and so of being superseded by another intellectual innovation. A certain initial enthusiasm is perhaps inevitable, but it must be tempered gradually if the new perspective is to avoid this fate. If it claims to illuminate a hitherto obscure pattern of thought and action, it must also be modest enough to recognize the legitimate claims of other ways of seeing the world.

The collection of essays in this volume has this dual purpose. On the one hand, we have attempted to illustrate the unifying elements in a series of diverse problems from the perspective of environmental concerns. Thus we have tried to show what such things as the bureaucratic structure of government, the market economy, public health, occupational safety, energy policy, and planning have in common when looked at from this perspective. In other words, we have constructed a paradigm in an attempt to determine whether by so doing we can improve our understanding of our present institutions and attitudes. It is an attempt to subject this perspective to a systematic test or trial, to see whether there is a basis for organizing differently our responses to social problems, and to see what the consequences might be of doing so.

On the other hand, these essays give due recognition to the existing patterns of attitudes and institutions, which respond to environmental concerns according to other interests and other ways of seeing the world. Neither our market economy nor our methods of public decision-making, for example, were 'designed' with environmental concerns in mind. Our political economy responds to those concerns on the basis of well-established private interests, distribution of power, lines of authority, and ideologies. Its 'instinctive' response to the expression of environmental concerns is to limit the definition of problems according to the capacity of existing institutions to deal with them – in this case, by extending the regulatory and price mechanisms already available to deal more adequately with pollution.

When a different way of looking at the world is first advanced, whether it is based on environmental or other concerns, some of its proponents argue that the existing order is fundamentally incompatible with the new perspective and must be dismantled. Even among these, however, there will be sharp disagreements over the design of what should replace it. They all confront those who believe that existing institutions can cope quite adequately with any specific problems. With respect to these positions there is probably as much disagreement among the authors in this collection as there is in Canadian society as a whole. On this point the reader is left free to form his or her own conclusions.

What unites the contributors here is the conviction that the perspective of environmental concerns is at least one important way of seeing the world today, and that adopting this outlook as part of a public decision-making process would improve the quality of life in Canadian society. Recent events encourage us to believe that this is not a quixotic or arbitrary stance, but one that is beginning to obtain public acceptance. The question of energy and industrial development corridors in the north is one of the most significant public decisions ever to be made in Canadian history. The special inquiry on environmental concerns (and the links between social and environmental issues) resulted in both a public information-gathering process and a report by Thomas Berger that are outstanding in quality and importance. Subsequently, the National Energy Board decision (July 1977) on pipeline proposals gave unusual prominence to both environmental and related social issues. I think Berger is correct in holding that the importance attached to such issues is a reflection of values widely shared in Canadian society.[3]

Like the inquiries just mentioned, the contributions in this volume focus on the interpenetration of social and environmental issues. For while it is true to say that 'environment' indicates a problem-domain that has a character of its own, and therefore can be isolated conceptually from other problem-domains (such as male-female or social class relationships), it is equally true to say that all environmental problems are simultaneously social and political problems. Our perceptions of the natural environment and the environmental impact of our activities are always conditioned by specific forms of social relations. The preceding essays illustrate some of the ways in which this interplay occurs in Canada today. In the following pages I wish to identify the general features that characterize the politics of environmental issues.

In general there are two kinds of environmental problems, quantitative and qualitative. In the first group I would place all those which we can isolate on the basis of investigations in the natural sciences and to which we can attach a quantitative measure. We must identify the toxic substances that circulate in the environment as a result of our producing and consuming activity, establish permissible levels of concentration for them, find ways of regulating those levels, and take remedial action where necessary. In addition we must undertake more complex ecological studies, involving the synergistic effects of combined biochemical impacts and the biological interactions among large numbers of species and inorganic materials, in small and large ecosystems. We must collect information and devise ways of regulating and mitigating impacts both for humans and for other species.

The qualitative problems are those arising out of our changing perceptions of the 'quality of life.' In part our judgments in this respect can be based upon scientific inquiry and quantitative data, for example in identifying medical problems associated with high noise levels, or in recording the flora and fauna that inhabit a marsh. But below the threshold of obvious physiological disturbance, how important to us is a reasonably quiet home environment, where it is still possible to identify the bird's call above the traffic din? How important is the local marsh, when the developer's bulldozers are poised on its periphery? How important is the habitat of an endangered species, when discoveries of mineral wealth are made there? Does it matter that the blue whale or a butterfly species becomes extinct? Environmental alterations change our conception of the quality of life in exceedingly subtle ways; and these 'intangible' matters are all too easily dismissed, when they appear to conflict with tangible benefits such as natural gas or new housing, as the snobbish concerns of privileged elites. Yet to me it appears odd to champion the cause of the less affluent while inferring that they have no interest in questions about the quality of life.

The difficulties in conceptualizing quality-of-life issues, particularly the distribution of benefits and costs associated with them, are the clearest indication that the political side of environmental concerns has two principal aspects as well. These I call the 'manifest' and the 'latent' politics of environmental issues. On the manifest level, environmental matters enter the social and political arena as one among many sets of private and public interests, jostling for attention with other determinants of well-being: love, wealth, income, status, power, authority, respect. They make up an intricate web of associations, and normally one of them is separated out only when something appears to upset the existing balance of factors that constitutes the individual or social definition of well-being. If any northern pipeline had been rejected on account of 'environmental' considerations, and if a natural gas shortage had occurred, would the adversities have affected everyone – or only the poorer groups? If there are higher costs due to mitigating environmental impacts when the pipeline is constructed, will the increased costs of fuel be apportioned equitably – or will lower income groups bear the brunt? Where there are disproportionate shares of income and wealth, as in Canadian society, public policies always have a differential effect on individual lives unless they are accompanied by offsetting measures (such as subsidies). This differential effect determines the responses to all social issues, including environmental concerns.

With respect to this manifest politics, environmental concerns will be – and must be – perceived as only one of the factors affecting the individual's well-being; and responses to them will be routed along the various avenues, established by the individual's position in the hierarchy of economic and political power, by which individuals seek to maintain and enhance their perceived well-being. With newly perceived issues of a very general nature (and environmental problems are of this nature), there is a strong tendency to define the issue as narrowly as possible. We find some security now in being able to restrict the definition of environmental problems to what can be expressed in quantitative terms, such as permissible levels in parts per million of lead in our backyards and of PCB s (polychlorinated biphenyl compounds) in our fish catch. For there is now a general consensus that we must have regulatory standards for obviously toxic substances and public responsibility for administering them – at least until they impinge upon economic interests, at which point we begin to shop around for the best bargain (that is, the least disturbing statistical evidence) in the scientific data.

The latent politics of environmental issues involves both the qualitative judgments mentioned above and also the incalculable risks we run in the massive environmental transformations, occurring everywhere on the globe, set in motion by modern industrial societies. We like to think that we are increasingly capable of controlling social and environmental change, through public policies that are shaped by scientific knowledge under democratic authority. Yet the very nature of a political economy founded on a complex industrial technology may defeat the possibility of effective control of public policy by informed citizens.

Our political economy has introduced, as a feature of everyday life, rapid change in our experiences and expectations, as well as in our social and natural environments. The prevailing 'climate' of our social surroundings, where such changes make themselves felt, plays a large part in conditioning individual behaviour and in affecting individuals' choices and preferences. Under circumstances of continuous change in our surroundings, are there sufficient points of stability on which individuals can base discriminating judgments? Or is our popular culture becoming a succession of lifestyle models, incorporating fleeting arrangements of preferences decked out on the roulette wheel of the marketplace? If much of our culture is made up of a quick turn-over of preferences, then the quality of our judgments and choices must begin to suffer, for discriminating judgments require some standard, some measure of value, against which alternatives are assessed. If this is indeed a widespread behaviour pattern, it will ultimately affect our political and public life. The sense of well-being will be influenced most strongly by whatever arrangement of political economy promises to ensure uninterrupted marketplace novelty – any novelty – and those environmental concerns that appear to threaten this program will be diminished accordingly.

A diminished sensitivity to environmental problems is a serious matter for a modern industrial economy, which pumps out this endless stream of marketplace novelties from its chemical wizardry. There is simply no time to do adequate environmental impact studies of the many thousands of new compounds (even if it were possible): we cannot wait for the new goods. The fluorocarbon propellants from aerosol sprays are already in the atmosphere, and the PCBs are in our water; we can only hope that whatever adverse effects might result will be manageable. Some persons in positions of responsibility think the Antarctic ice mass would be an ideal dump site for radioactive wastes from nuclear plants; others oppose this, and one doesn't yet know who will have the final say.

How could we have known, when as sovereign consumers we expressed our preference for pressurized cans over manual-pump spray containers, that the gases that saved us the exertions of our fingers rose through the atmosphere to react with ozone? (What is ozone anyway?) How could we have known, when inserting carbon paper into our typewriters or switching on fluorescent lights, that later, decomposing in our garbage, they would allow polychlorinated biphenyl compounds (what?) to seep into our rivers and lakes, and to become more highly concentrated along the food chain, until the reproductive rate of herring gulls and double-breasted cormorants dropped sharply and until governments advised 'all women capable of pregnancy' (at least those who read government bulletins) not to eat any fish brought home from their husbands' outings on Lake Ontario? Try to imagine the kind of bureaucratic apparatus that would be required adequately to enforce environ-

mental standards for thousands of potentially harmful chemicals –
studying their effects, monitoring their movement through ecosystems
and species, regulating their use in industrial plants, informing the public
about hazards, compensating individuals and businesses for losses, and
funding medical remedies for exotic complaints.

The latent politics of environmental issues is defined most sharply by
the case of environmental standards. By 1975 the United States Food
and Drug Administration had set a limit of five parts per million for PCB
concentrations in fish for human consumption (at that time Canada had
no regulations on the matter).[4] The International Joint Commission, a
US-Canadian advisory agency on water bodies, proposed a limit of one
part in ten million; at the same time, the US Environmental Protection
Agency was studying a proposal to set the limit at one part per trillion –
effectively a zero limit and an indication of just how hazardous PCBs
might be. There seemed to be much uncertainty on scientific grounds
about what standard was necessary for protecting human health.

But the Great Lakes have a considerable commercial and sports
fishing industry, which was already suffering losses due to regulations
on other chemical contaminants. A US official told a reporter that technical experts were asked for 'a number that would not badly impact
economic interests' during discussions on setting more stringent standards for PCBs. This theme recurs constantly in published reports about
changes in environmental standards – for asbestos, lead, radioactive
materials, and many others. The political process mediates temporary
trade-offs among a wide range of considerations, including scientific
research, unemployment, capital investment, public awareness, and
rates of occupational disease. The 'number' selected for an environmental standard only appears to be derived directly from the pure
disinterested inquiries of the laboratory; in fact, it usually represents a
rough compromise among vested interests, balancing science, politics,
and economy on the knife-edge of potential catastrophe.

So far as the manifest politics of environmental issues is concerned,
environmental hazards are one among many factors in the conflicting
play of social interests. Accepting this was bitter fare for those who
imagined, in the first flush of excitement over the discovery of
'spaceship earth,' that the magic term *ecology* would still the endemic
clamour over how to parcel out the planet's booty. It is instructive to
review how various currents converged to form a heightened sensitivity
to environmental concerns, why inevitably this new sensitivity was

domesticated by established institutional processes, and finally what was represented in the changed alignment of interests.

Many divergent sources came together in the 1960s to turn environmental problems into a major social issue. I have isolated five of these below, but there were also others. After describing each briefly, I shall discuss various institutional responses to this issue.

First, environment was simply a new label for some very old problems in industrial society. The horrible effects of unhealthy working conditions in mines and factories are the most persistent of these; and despite significant improvements over the years, largely forced onto the statute books by labour struggles, this remains one of the worst blights on our social record. Even the very diagnosis of hazards still entails a battle against bureaucratic resistance, including the pitiful spectacle of attempts by workers' widows to marshal medical evidence of occupationally caused deaths before sceptical compensation boards. Medical science has neglected occupational disease, with the result that reliable statistics rarely exist. Yet even if they were available, such quantitative measures could convey not even the slightest impression of the individual suffering represented in them: those who doubt this should consult Elliott Leyton's *Dying Hard*, which records the slow agonizing deaths in their families' presence of Newfoundland men poisoned in the fluorspar mines. 'Simple' things like noise, dust, and heat take their toll along with asbestos, vinyl chloride, and lead, as the hidden human cost of profit and industrial wealth.

Second, the new label for these old problems was useful in helping us to see particular outrages as part of a *pattern* of abuse inflicted on people and other living entities in our environment. This includes deleterious impacts on the population as a whole, as well as hazards peculiar to certain occupational settings. Air and water pollution are the most widely recognized cases. The larger picture aided us in understanding the environmental context of diseases such as cancer; at the same time we have achieved greater insight into connections among particular health problems that are related to general factors, such as stress, in our social environment. Any hope of alleviating the enormous personal and financial costs incurred as a result seems to depend, at least in part, on applying a unified environmental perspective to a program of preventative medical care.

Third, our heightened sensitivity permitted us to detect or better appreciate some hitherto unknown or obscure matters. We learned much more about the increasing concentration of toxic chemicals through the various stages in the food chain (levels of chlorinated compounds are 40,000 times higher in fish than in the waters they inhabit, and the concentrations increase in the birds and polar bears that in turn feed on them). Research in atmospheric chemistry led to controversies over aerosol sprays, supersonic aircraft, and the shrinking of the Amazon forest. Dedicated followers of these episodes discovered that vast quantities of antibiotics are routinely administered to animals raised for human consumption, resulting eventually – it seems likely – in increased resistance to drug treatment of disease organisms. And so on. After digesting a certain amount of such information one reaches a threshold limit, and a mildly catatonic state ensues. So the knowledge that one's lowly beer has been treated with twenty-five chemical additives (including one to inhibit foaming when it is uncapped and another to produce foaming when it is poured into the glass) no longer registers, and one orders another, which seems to help matters considerably.

Fourth, some existing problems were exacerbated by the greater attention bestowed on these concerns. The more indiscriminate attacks launched by environmental groups on pollution or on energy-generation facilities have been met with accusations that they were indifferent to the plight of the poor and the unemployed, who would be most severely affected by higher prices and construction delays. The ritualized burial of a brand-new automobile was furiously denounced, and independent California loggers expressed their displeasure over filling out environmental impact assessment forms for cutting trees in no uncertain terms. In international relations the Stockholm Conference erupted in accusations that environmental protection was only the latest stratagem in the industrialized world's plot to ensure that the developing nations remained only raw-material suppliers. The exporting of pollution, for example the 'acid rain' that Britain sends to Norway, has not yet caused serious confrontations, although it may well do so in the future.

Finally, there was a significant reordering of priorities with respect to issues of social change among many individuals. After the first wave of initial enthusiasm passed, many persons remained dedicated to long-range efforts designed to deepen our understanding of environmental problems and to enhance the capacity of our institutions to deal with them effectively. Campaigners on behalf of whales and seals can receive wide publicity, leading to accusations of theatrical posturing. But thousands also work more quietly for governments, public interest groups, corporations, and universities; popular and academic literature, television, film, and other media have communicated the results with increasing effectiveness, and at the very least we seem to have now a public resolve to include environmental considerations in all of our important policy choices.

In these dimensions of an emerging issue there are, I believe, sufficient grounds for asserting that more than a passing fad is at stake. From now on, I do not think that we shall ever (except perhaps under extreme social stress) ignore environmental impacts in deciding major policy questions; if this is indeed what we do, we shall have witnessed the beginnings of a significant change in our social consciousness. I am not claiming that the main features of our political economy and social relations have been altered thereby. It was both inevitable and necessary that our political economy would respond to this emerging issue by seeking to contain the understanding of environmental concerns within its own limits. These limits are the capacity of existing institutions to manage problems in such a way that the general alignment of established social interests – such as the distribution of wealth and power – is not unduly threatened.

Setting aside speculative forecasts about how significantly environment concerns will figure in the future, we can ask what the institutional responses to this issue tell us about those institutions themselves.

Specialists in comparative politics have begun to study such responses as an index of variants in political systems.[5] By comparing institutional adjustments to pollution abatement programs in the United States, the Soviet Union, and Japan, for example, we find new ways of discovering which interest groups or professions can make their views felt in influencing policy decisions. In addition, we can discover the extent to which bureaucratic agencies adjust both to new information and to pressures brought to bear on them by citizens' initiatives. During 1977 there was considerable public controversy over nuclear energy development in many Western nations (East Germany, on the other hand, simply announced that this was a non-issue and that nuclear power would assist the continued march of socialism). In Great Britain and Canada official inquiries were held at which citizens' groups participated extensively, and it is clear that the contributions of public interest organizations and concerned individuals are being taken into account in a serious way. However, no such opportunities were presented in France or West Germany; the violent public demonstrations in both countries are at least in part due to the government's failure to provide any other avenue for the expression of dissent against established policies.

In economic affairs the path of least resistance is to adjust market

forces to deal with more stringent pollution standards and related requirements. The additional expense of better anti-pollution equipment is
therefore included in the costs of production, sometimes resulting in
higher prices for the consumer. But there are other, less obvious
changes as well. Adjustments to more elaborate environmental impact
regulations and to capital investment expenditures for pollution abatement equipment may be easier for large corporations than for small- or
medium-size firms. Thus one unintended consequence of allowing market forces to govern the economic effects of improved environmental
standards may be to augment the already considerable thrust towards
greater concentration of economic power. At the same time, the
world-wide scope of the multinational corporation's activities serves as
a check on the ability of national governments to enforce stricter environmental regulations. When Canadian provinces lowered permissible levels of concentration in plants processing asbestos, the corporations transferred the 'dirtier' operations to those Third World countries
whose compliant authorities were willing to trade off health hazards for
economic benefits.

Threats to employment and capital investment levels in local areas
have been cited in almost every case where environmental concerns
have been raised about large-scale manufacturing, mining, or energy-
production facilities. Labour union leaders and members have usually
(but not always) stood shoulder-to-shoulder with corporation executives in dismissing or in seeking to minimize such concerns. More
recently labour unions have disagreed among themselves over these
matters. In Great Britain the leader of the Yorkshire coal miners, appearing at a public inquiry into the safety of fast-breeder nuclear reactors, argued that these reactors present an extreme environmental
hazard; Britain could avoid developing them, he suggested, and instead
rely on increased coal supplies for generating electricity until solar
energy could make a significant contribution to our energy needs in the
coming century. He was vigorously opposed by the union representing
nuclear station employees, which denied that these generating plants
were environmentally hazardous – and emphasized the occupational
disease and accident records of coal miners.

Finally, how have governments reacted to the new social issue? Here
one must admire the elegant simplicity of the initial response: they
create a 'Ministry (Department) of the Environment.' Like good dance
troupes, experienced governments can come up with a new act on very
short notice. When they perceive the situation as only a matter of riding

out a storm, this bureaucratic choreography can be quite amusing. One Italian government created and then dismantled a department of the environment in a mere eight months. In Canada the federal department sometimes does solos and sometimes is paired with Fisheries, which is perhaps a less inspired union than the provincial ministries that join 'culture' with sport or recreation. In Britain the secretary of state for the environment is responsible for ancient monuments and historic buildings, driver licence examinations, town planning, and parks, as well as for air and water quality and toxic chemicals.

Certainly the effectiveness of such bureaucratic conglomerates has been impaired by the rather hasty manner in which older regulatory bodies, such as water resources boards, were patched together under a rubric that seemed more appropriate for present-day concerns. Housing various branches in the same office tower creates the illusion of a co-ordinated approach to environmental problems, but this co-ordination will require some time to show itself effective in practice. The illusion is compounded by noble statutes such as Ontario's 1975 Environmental Assessment Act, which potentially subjects every impingement on the environment (that is, every human action) to the watchful eye of government; its application, of course, is confined to a rather more narrow range of problems both by dictates of common sense and by the substantial interests of other ministries. Yet it is certainly useful, from a public relations standpoint, for governments to claim that a designated line of responsibility for environmental matters has been established.

I have offered only a few illustrations of institutional responses to environmental issues in four areas (political systems, business, labour organizations, and government administration). Two different approaches have been indicated. On the one hand, we can use the perspective of environmental concerns to shed light on the internal dynamics of these institutions, without necessarily having any special interest in environmental concerns themselves. On the other hand, with regard to the intrinsic nature of such concerns – what should be the permitted levels of toxic substances in air and water, for example – we can assess the capacity of our existing institutions to respond in a manner that seems appropriate to the gravity of the problem as we now understand it.

Institutions respond to issues (or ignore them) by balancing two kinds of pressures: those emanating from what we might call their 'deep structure,' which represents the historical sediment of the most influential social class and ideological forces; and expressions of perceived self-interest by individuals and groups in the present which seek to influence social policies. The degree of flexibility in institutions, that is, the extent to which they can take at least some effective action in confronting new issues, varies considerably from one society to another. Responses to environmental issues in North America reveal both the extent and the limits of their adaptability. Shortly after the potential danger of fluorocarbon propellants was reported, manufacturers offered manual-pump spray containers once again, and in advertising them they pointed out how much more economical they were for the consumer! On the other hand, the balancing of pressures with respect to mercury pollution in northern Ontario rivers has been far less imaginative, to say the least. Providing a dole of frozen fish to native peoples as a substitute for banned river fish shows a gross insensitivity to the wider cultural significance of food-gathering activities. The mere substitution of one quantity of fish by another, while ignoring the qualitative differences both in the fish and in the native culture's food-gathering activities is a depressing example of the way a welfare-industrial society can reduce the quality of experience to simple numerical equivalents.

Having looked at institutional contexts, we must now turn briefly to the consideration of reactions at the individual level. Apart from those who have a professional interest in these matters, what is the meaning of environmental issues for the busy citizen in an industrial society? Environmental problems constitute a small proportion of the news items that flash by every day; major oil spills make headlines for a day or two, but most such problems (such as the passage of chlorinated compounds through the aquatic food chain) are quite complex in nature and require serious and sustained attention if they are to be understood. There are real difficulties involved in attempting to incorporate our knowledge of them, together with a wide range of other considerations, into changed patterns of attitudes and behaviour.

Thus most persons will be content to trust institutionalized scientific research and regulatory authority to manage environmental problems as they occur, without even becoming aware of them. Their real impact will be felt only indirectly in our everyday affairs at work and shopping. Greater expenses for pollution control equipment will be passed along by producers and will appear as higher prices for goods and services; some of them, such as more expensive emission control devices for automobiles, are quite evidently the outcome of environmental policies. (The widespread practice of sabotaging the apparatus to improve gas mileage reveals the gap between individual and social benefit-cost ratios.) Impact assessment requirements for large-scale developments can cause construction delays and threaten short-term employment prospects. More stringent standards for hazardous substances may affect decisions on siting new mining and manufacturing plants, an especially serious matter for economically depressed areas with persistently high unemployment levels. Temporary bans on fishing, controversies over whether to spray New Brunswick forests against the spruce budworm, measures to reduce dust and gases in mines: all of these and many other questions have been accompanied by threats from large corporations to shut down operations that would, it is claimed, become 'uneconomic' because of higher costs.

What are the individuals whose present livelihood is at stake to do under these circumstances? Invariably the spokesmen for the dominant economic interests maintain that the 'alleged' environmental hazards have been either wildly exaggerated or completely unproved. Contradictory evidence from different scientific researchers is advanced. In many cases the problems, such as air pollution from automobiles, are spread over a wide area, but the economic impact of mitigating them is concentrated in certain localities (Detroit or Oshawa). Sometimes the problem is exported entirely to another country, as is the pollution of Manitoba water by North Dakota irrigation schemes or the high salinity of Colorado River water when it crosses the Mexico border. Individuals face plenty of threats – poverty, discrimination, family breakup, illness; and, on the other hand, plenty of opportunities – financial success, prestige and power, comfort, security. These threats and opportunities arise directly and immediately in everyday life. Most environmental concerns appear ambiguous, remote, or technical (when they are known at all), except to the relatively few persons whose lives are now seriously affected by them.

At present the benefits of more rigorous environmental standards for most people tend to be confined to intangible quality-of-life measures, whereas the costs are expressed in far more tangible terms such as jobs. In a society where the sense of well-being and self-worth is largely defined by the level of personal consumption of goods and services, there can be little doubt as to which is the more compelling consideration. I suggested earlier, however, that Thomas Berger's view of the accepted place for environmental values in Canadian culture cannot be dismissed out of hand. These values encompass not only the preservation of natural wildlife and wilderness but also a growing determination to reorient social priorities in some of the other areas discussed above.

During the past few years great advances have been made in occupational health and safety legislation, in large part because individual workers and unions decided to assign far more importance to such matters than was done in the past. Obviously much remains to be accomplished. But when the benefits begin to be realized, similar initiatives in other areas will be encouraged.

A sense of improvement or deterioration in the quality of life is often elusive and hard to measure. It is linked with the feeling that the relation between the individual and his or her surrounding environment has been enhanced or degraded. It is usually experienced indirectly rather than directly, as the contextual background (encompassing both physical environment and social relations) of daily activities; on it depends in large measure the degree of enjoyment derived from those activities.

Major changes in this contextual background affect our ranking of preferences and personal objectives, our awareness of sensations in our environment (sounds, colours, smells), and therefore the quality of the satisfactions we experience in work, play, and consuming. In transitional periods when, for example, many people are moving from rural to urban environments, such contextual changes can lead to a considerable degree of ambiguity in personal experience. In our society one indicator of this ambiguity is the widespread use of background imagery derived from rural and wilderness settings in advertising; these settings, I think, tap older reservoirs of feeling through which individuals interpret the qualities of their new experiences, which are now oriented primarily towards the omnipresent consumer-goods displays in urban settings.

In the preceding section I tried to show how the manifest politics of environmental issues emerges from the tension between established institutional forces and individual perceptions of well-being and self-interest. The resolution of this tension determines how environmental concerns will be understood and how they will be ranked in the range of current social issues at any moment. But if it is correct to say that changes in the contextual background of individual experiences affect in turn these individuals' perceptions and judgments, then one must look beyond the level of manifest politics. This is not easy, for the circularity of the process – contextual changes influencing individual preferences, resulting eventually in changed perceptions of the background context itself – is an especially difficult case for social analysis.

One way of explaining this 'contextual circle' is as follows. The so-called mass consumption society that emerged in North America following the Second World War created not only greater quantities of consumer goods but also a new 'social world.' Individuals have been encouraged to abandon long-established practices (principally thrift, saving, careful maintenance and re-use of personal possessions, home production of such things as food and clothing) and to adopt different ones: rapid turn-over of possessions, waste, credit buying, constant attention to the stimulus of the marketplace, denigration of 'home made' items. I wish to suggest, without developing the point in further detail here, that this changed general orientation affects the individual's sense of enjoyment, satisfaction, and well-being – in short, the ingredients that make up that elusive indicator known as the quality of life.

The subtle transformations in quality-of-life criteria, flowing from the interplay of changing preferences and their background context, are a latent or hidden dimension of politics. These transformations are present implicitly, rather than explicitly, in the political choices by which we influence the direction of public policy. Because quality-of-life criteria are constituted, at least in one important sense, in the relation between individuals and their environments, the substantive issues raised by environmental problems are a major component of the 'contextual circle.' This is the concealed dimension of a behaviour pattern whose more visible manifestation is our willingness to remodel the physical environment and to ransack its mineral resources at a reckless pace. In it simmers the latent politics of environmental issues.

One feature of this latent politics can be indicated by extrapolation from what is already partially apparent. This involves the probable long-run consequences of attempting to manage the environmental risks of our industrial economy's chemical wizardry. The other is more speculative and involves the possible consequences of failing to re-evaluate the environmental values that are now shaping our quality-of-life criteria. I shall consider briefly each of these in turn.

Modern industrial societies all have elaborate bureaucracies through which are co-ordinated, with varying degrees of efficiency, the innumerable daily production and consumption transactions of its citizens. Some see in them a rational division of labour and a sensible decision-making forum; others regard them as merely instruments for social control by elites whose hierarchical structure stifles personal initiative. But not even their harshest critics could maintain, it seems to me, that at current levels of activity a national economy in an industrialized society, with its vast network of domestic and international exchanges, could be managed without elaborate institutional structures of some kind.

There is considerable ambivalence about large bureaucracies (private and public) at the popular level. Attacking them is one of the simplest ways for a political candidate to excite an audience during an otherwise dull campaign speech. Since the passion subsides quickly thereafter, the electorate appears to anticipate that nothing much will change – evidence perhaps of their realization that dismantling them would disrupt the entire fabric of daily life and that the price to be paid might be too high.

No one can doubt that there is a real issue at stake, namely the difficulties of exercising some measure of democratic control, through informed opinion and accountability in public policy, over administrative units. As public policy issues become more complex, reflecting the larger scale of socio-economic undertakings, decision-making mechanisms come under greater stress. The stakes for everyone are very high in projects such as James Bay and northern pipelines. The environmental impact of large-scale energy developments is the clearest example of why environmental problems in general *may* accelerate the tendencies in our society towards more centralized decision-making.

The environmental hazards associated with nuclear technology, for example, arise with respect to both present uses of toxic substances and radioactive waste management over very long periods in the future. Thus they create 'stress' for our social institutions today and also bequeath the strain to succeeding generations. This is characteristic of many other environmental problems associated with toxic chemicals. There is no known way of recovering substances such as PCB s once they are dispersed in the air, water, and soil, and apparently we shall have to monitor their effects indefinitely. A specific illustration is the 300,000 tons of arsenic waste (a by-product of processing gold) now stored in abandoned gold-mine shafts near Yellowknife, which may leach into surrounding water bodies should permafrost thawing occur: 'The Yellowknife reports said that the waste does not present an immediate problem as long as it is watched carefully. However, there were fears that when the mine eventually closes, it will flood. Flood waters would melt the permafrost and the arsenic might seep out into Great Slave Lake, a huge body of water connected with major northern river systems. The Environment Department is said to have recommended that the temperature of the permafrost be checked constantly and that the storage shafts be pumped forever so no flooding occurs. A Northern Affairs official in Yellowknife was quoted as saying that checks were to have been made from time to time but the federal Treasury Board never had approved a budget for the project.'⁶ What kinds of guarantees can be given that the shafts will be pumped 'forever'? And how many other, similar situations are there which should be 'watched carefully'?

We still have no idea even how many substances present serious hazards or what the permissible concentrations for them should be. In 1975 the US government noted that one to two thousand chemicals used in industry had been identified as being potentially harmful and that five to six hundred new substances were being introduced each year; by 1975 standards had been established for only twenty-five. Yet setting a standard is only the first step. Monitoring devices have to be developed and put into place, processes for taking readings devised, enforcement practices implemented, medical examinations conducted, and appropriate treatment offered. All of this must be done not only in large plants, where concentrations of workers could make mass monitoring techniques relatively easy to administer once they were in place, but in almost every work setting. A three-year study in the northwestern United States revealed that 'one out of every four workers in a sample of small business has an occupationally derived disease and 89 per cent of those are not reported as required to the Labour Department.'⁷

At present, these are not matters of primary social concern. But the massive environmental impacts of industrial economies will become increasingly evident, and if we continue to ignore or downplay the 'unintended' side-effects, we will do so at our peril. Our squandering of resources and energy and our dependence on the novelties produced by industrial chemistry is what creates the 'need' for nuclear power and the risks attendant upon both it and the other hazardous contaminants circulating in the environment. The magnitude of the risks remains largely hidden from us on account of the complex ecological interactions through which they make themselves felt. As a result, the potential political tensions in this issue remain hidden as well.

Environmental problems will set some difficult tests for our political institutions. What will make these tests especially hard for us is the fact that we have come to define our environmental values primarily in relation to demands for steady economic growth – or, more precisely, in relation to a sense of well-being that seems to require, apparently forever, a regular increase in the GNP. We have only begun to take the first steps towards understanding, on a theoretical level, why this objective is ultimately self-defeating for individuals, why each new level of consumption yields little lasting personal satisfaction but rather only stimulates further expectations and aspirations.⁸ We do not know what are the possibilities that significant numbers of people might turn away from this behaviour pattern, or what developments might give rise to a different one that did not tie the sense of well-being so exclusively to consumer aspirations. We have only a few intimations of what are the political implications inherent in a different pattern, for example, a 'steady-state' or 'conserver' society.[9]

The existing behaviour pattern encourages us to regard the natural environment from a purely instrumentalist perspective, that is, to regard it solely as a means for supplying our wants. We see little intrinsic value in a relatively undisturbed natural landscape – its physical contours as well as the flora and fauna inhabiting it – and we undertake as quickly as possible whatever modifications are required in order to make it yield those produced things we now prize so highly.

We recall sometimes that other people (the 'backward' or 'primitive' sort) took a different view. They also drew sustenance from their natural surroundings; but to them it had meaning or symbolic significance over and above its ability to supply their wants. The land itself, its particular features, the flora and fauna it nourished, provided a basis of continuity for human cultures. It was what had 'always' been there and what they hoped to bequeath with undiminished abundance to succeeding generations. The animals especially were more than a resource: they too possessed 'spirits' and in this common dimension helped sustain the bonds of identification between man and his habitat that gave meaning to existence. In the nineteenth century Chief Joseph of the Nez Percé told the European invaders on this continent who were indiscriminately slaughtering the bison herds that 'without the animals man would perish of a great loneliness of spirit.' The invaders were unpersuaded.

Thomas Berger's pipeline inquiry report presented to us, in the context of a major social policy issue, a challenge based in part on a reaffirmation of that older lifestyle. He investigated not only the more immediate issues of industrial development and native settlements in the north, such as alcoholism and family disintegration, but also the deeper structure of experience that has its source in a special orientation to the natural environment. He quotes a native witness: 'Being an Indian means being able to understand and live with this world in a very special way. It means living with the land, with the animals, with the birds and fish, as though they were your sisters and brothers. It means saying the land is an old friend and an old friend your father knew, your grandfather knew, indeed your people always have known.'[10] And he recommended that we southern Canadians take these expressions seriously when we decide how to go about introducing further industrial development in the north.

Many persons might regard this sentiment with cynical disdain as a manoeuvre by native peoples to improve their bargaining position. Cash settlements for native land claims, on the other hand, have the virtues of simplicity and tidiness – and even more so the advantage of specifying terms that are readily comprehended by us. Despite the fact that Berger made it quite clear he was not urging us to adopt for ourselves the native peoples' attitude towards the natural environment, its presence alone as a factor in policy deliberations is unsettling.

It *could* also represent an opportunity for us, if not now then perhaps sometime in the future. For eventually we may re-examine our existing behaviour pattern and ask why, having raised our industrial productivity yet another few notches, we are still left with pervasive economic insecurity, tensions stemming from the unending scramble for status, unfulfilled wants that always exceed incomes – in short, without a clear sense of social and personal well-being.

I am not suggesting that we can find in the lifestyle of native peoples an alternative model for ourselves. We shall have to create one that is based upon our own cultural history. I think that a reorientation of our environmental values is an important ingredient in any attempt to do so. What we can learn from the human history we share with other cultures is that some caring attitude towards the natural environment, some feeling of belongingness within the natural order, is one essential aspect of the sense of well-being and personal satisfaction.

I do not think there would be much disagreement with the statement that most persons derive a deep sense of satisfaction from caring for others, and from being cared for in turn by them. This is expressed in our philosophical and religious tradition by the doctrine that a person should not be treated by others merely as a means to the realization of their objectives, but always also as a being of intrinsic worth, as worthy of care and respect simply because he or she is part of the community of persons. Of course persons are treated as means for the satisfaction of others, for extracting love, profit, and innumerable other resources. The traditional doctrine recognizes this but also teaches us to bridle our extractive impulses by reflecting on whether or not what we require of others on our own behalf is good for them as well. And our society today acknowledges, on the level of public responsibility, a degree of mutual obligation for the welfare of all persons far higher than what existed in the past.

Earlier cultures included both the land and other animals, as well as humans, in the 'community of beings' that formed the frame of reference for their sense of shared concern and responsibility. Our industrial culture, on the other hand, influenced by the idea that we have 'conquered' nature, discourages us from following their example. Indeed we cannot do so: our entire way of life is far too different. However, we should not take this to mean that we need not care at all about our relationship to the natural environment. Today we do see the environment almost exclusively as a means for providing resources for the satisfaction of our needs. But we should not forget that out of those resources we fashion the material goods that we use, and that when we select from among them those that appeal to us, we are choosing those which we think express our own identity and personality, those which reflect in part the qualities of our selves that we would like others to regard as worthy of respect. The connection between our concern for our own well-being and that of others, and our relation to the natural environment, is not an immediate one (as it is in earlier cultures); it is an indirect one, and it is concealed by the other connection we have made so important in our lives, namely that between the sense of well-being and the increase of goods.

Should we attempt to devise a more direct connection between our sense of well-being – or our perception of the quality of life – and our relationship to the natural environment? Does it make sense, in terms appropriate to our industrial culture (and not in terms of the conceptions in earlier cultures), to speak of broadening our caring attitude to include the land and other natural entities as well as persons? Raising such questions carries no implication that the environment can or should cease to have a vital significance to us in economic terms, as the material basis for the satisfaction of needs. Of course it must continue to do so. But can it have a much greater non-economic significance as well?

Much depends upon what choices we make with regard to future social and economic development in general. Earlier in this essay I referred to a number of studies that suggest we cannot expect economic growth in itself to supply a better sense of well-being and personal satisfaction. If these arguments are correct, we shall have to search for other routes. I believe that the experience of caring and being cared for by others is one of the most stable and enduring foundations for the sense of personal worth, dignity, and self-respect, and thus of the sense of well-being. Broadening and deepening that caring attitude to include the land and other natural entities, whereby they too would be regarded not only as mere means for the satisfaction of our needs but also as things intrinsically worthy of respect and care, might be one among several new ways we could find to enhance our well-being.[11]

A reorientation of our social life that made market-based aspirations less important to us than they now are would release new kinds of political tensions. It is impossible to say at this time whether these new tensions would be more constructive in the long run than are those we experience today. This aspect of the latent politics of environmental issues cannot yet be brought to light.

NOTES

1 Commoner, *The Poverty of Power* (New York 1976), 216
2 A. Downs, 'Up and Down with Ecology – the "Issue-Attention Cycle," ' *Public Interest*, no 28 (Summer 1972), 38–50
3 Berger, *Northern Frontier, Northern Homeland: The Report of the Mackenzie Valley Pipeline Inquiry* (Ottawa 1977), I, 30
4 *Globe and Mail*, Toronto, 4, 6, 7 Oct. 1975
5 C.H. Enloe, *The Politics of Pollution in a Comparative Perspective* (New York 1975); D.R. Kelley *et al.*, *The Economic Superpowers and the Environment* (San Francisco 1976). Other related works published recently are H. Stretton, *Capitalism, Socialism, and the Environment* (London 1976); G. Hardin and J. Baden, eds., *Managing the Commons* (San Francisco 1977); M.F. Mohtadi, ed., *Man and His Environment* (London 1976), II.
6 *Globe and Mail*, 9 Feb. 1977
7 *Ibid.*, 28 Jan. and 28 April 1975
8 T. Scitovsky, *The Joyless Economy* (New York 1976); F. Hirsch, *Social Limits to Growth* (Cambridge, Mass. 1976); W. Leiss, *The Limits to Satisfaction* (Toronto 1976)
9 Science Council of Canada, *Canada as a Conserver Society* (Ottawa 1977); W. Ophuls, *Ecology and the Politics of Scarcity* (San Francisco 1977). The best account is C. Taylor, 'The Politics of the Steady State,' in A. Rotstein, ed., *Beyond Industrial Growth* (Toronto 1976).
10 Berger, *Northern Frontier*, 94
11 For a fuller discussion, see W. Leiss, 'Dominion over Nature and Respect for Nature,' in V. Mathieu and P. Rossi, eds., *Scientific Culture in the Contemporary World* (Paris, forthcoming 1979).

Contributors

Grahame Beakhust is a member of the Faculty of Environmental Studies, York University

Mario E. Carvalho is a professor in the Department of City Planning and associate dean of the Faculty of Architecture, University of Manitoba; his areas of specialty are regional planning and planning theory

C.A. Hooker is professor of philosophy and environmental engineering at the University of Western Ontario

Reg Lang is a professor in environmental studies at York University; he has extensive experience in urban/regional planning, environmental planning and management, and engineering as a civil servant, consultant, and researcher across Canada

C. Clifford Lax is a Toronto litigation lawyer with a specific interest in environmental law; he is a past president of the Canadian Environmental Law Association and the Canadian Environmental Law Research Foundation, and is at present chairman of the Environmental Section of the Canadian Bar Association

William Leiss is a professor of environmental studies and political science at York University, and author of *The Domination of Nature* and *The Limits to Satisfaction*

Robert Macdonald is an associate professor in environmental studies at York University; he holds a PH D in physics from the University of Toronto

Robert Paehlke is associate professor of political studies at Trent University and founding editor of the journal *Alternatives: Perspectives on Society and Environment*

John E. Page is a professor in the Faculty of Environmental Studies, York University, whose special areas of interest are urban-regional planning and man-environment relationships; he holds his PH D from the University of Pennsylvania

Robert Sass has been an associate professor of administration, University of Regina, and is at present director of the Occupational Health and Safety Division and associate deputy minister in the Saskatchewan Department of Labour

Mark E. Taylor studied zoology and botany at London University and obtained his PH D from the University of Toronto; he was an assistant professor at Pahlavi University, Shiraz, Iran, and at York University, and currently teaches courses in ecology, vertebrate morphology, and human biology at Cariboo College

Robert van Hulst teaches biology at Bishop's University, Lennoxville, Quebec

Peter A. Victor is a research consultant with Middleton Associates, associate professor in environmental studies (part-time) at York University, and a research fellow at the Institute of Environmental Studies, University of Toronto